A Photographic Atlas for the Anatomy and Physiology Laboratory

FIFTH EDITION

Kent M. Van De Graaff

Weber State University

John L. Crawley

Morton Publishing Company

925 W. Kenyon, Unit 12
Englewood, Colorado 80110

To our families for their sacrifice, support, and love.

Table of Contents

A Photographic Atlas for the Anatomy and Physiology Laboratory

Preface

Human anatomy is the scientific discipline that investigates the structure of the body and human physiology is the scientific discipline that investigates how body structures function. These subjects may be taught independent of each other in separate courses, or they may be taught together in integrated anatomy and physiology courses. Regardless of whether or not anatomy is taught independently from physiology or if the two disciplines are integrated as a single course, it is necessary for a student to have a conceptualized visualization of body structure and a knowledge of it's basic descriptive anatomical terminology in order to understand how the body functions.

A Photographic Atlas for the Anatomy and Physiology Laboratory is designed for all students taking separate or integrated courses in human anatomy and physiology. This atlas can accompany and will augment any human anatomy, human physiology, or combined human anatomy and physiology textbook. It is designed to be of particular value to students in a laboratory situation and could either accompany a laboratory manual or in certain courses, serve as the laboratory manual.

Anatomy and physiology are visually oriented sciences. Great care has gone into the preparation of this photographic atlas to provide students with a complete set of photographs for each of the human body systems. Human cadavers have been carefully dissected and photographs taken that clearly depict each of the principal organs from each of the body systems. Cat dissection, fetal pig dissection, and rat dissection are also included for those students who have the opportunity to do similar dissection as part of their laboratory requirement. In addition, photographs of a sheep heart dissection are also included.

A visual balance is achieved in this atlas between the various levels available to observe the structure of the body. Microscopic anatomy is presented by photomicrographs at the light microscope level and electron micrography from scanning and transmission electron microscopy. Carefully selected photographs are used throughout the atlas to provide a balanced perspective of the gross anatomy. At the request of several professors who used previous editions of the atlas, the muscular and circulatory sections have been expanded and improved with new photographs, illustrations, and tables. The section on articulations has been improved with the inclusion of photographs of joint dissections. Selected X-rays, CT scans, and MR images depict structures from living persons and thus provide an applied dimension to the atlas. Great care has been taken to construct completely labeled, informative figures that are depicted clearly and accurately. The terminology used in this atlas are those that are approved and recommended by the Basle Nomina Anatomica (BNA).

Preface to Fifth Edition

New editions are desirable for authors because it presents an opportunity to improve upon a successful product. An extensive revision, such as is presented in the fifth edition of A Photographic Atlas for the Human Anatomy and Physiology Laboratory, requires an inordinate amount of planning, organization, and work. As authors we have the opportunity and obligation to listen to the critiques and suggestions from students and faculty who have used this atlas. This constructive input has resulted in a product that is greatly improved. We appreciate those who have taken the time to provide suggestions and indicate corrections.

One of the objectives in preparing this atlas was to create an inviting pedagogy. The page layout has been improved by careful selection of photographs, and when necessary, provide accompanying line art. The tables were redesigned to enhance the retrieval of information. Each image in this atlas has been carefully evaluated for its quality, effectiveness, and accuracy. White backgrounds for the depicted specimens enhance the clarity of the images and tend to provide an open and inviting appearance to the pages. Most of photographs have been improved or replaced by better photographs and the leader lines are better spaced to aid in the identification of structures. Major changes were made in chapters devoted to the muscular system, nervous system, and sensory organs. Human cadavers were dissected to obtain better quality photographs of the muscular system. Quality photographs of detailed hand and foot dissections add to the value of this edition. Sixteen additional pages were added to this edition, thus enabling enlarged images in certain chapters and additional photographs in other chapters. Even the surface anatomy photographs have been replaced enabling features to be more readily identifiable. New cat dissections enhance the quality of this atlas. At the request of several users of A Photographic Atlas for the Human Anatomy and Physiology Laboratory a chapter 21 was added depicting the anatomy of a rat. Some instructors use preserved rats as mammalian models in their anatomy laboratories.

Acknowledgments

Many individuals contributed to the preparation of the fifth edition of A Photographic Atlas for the Human Anatomy and Physiology Laboratory. We are especially appreciative of Amber M. Hughes, Jon S. Williams, and Aaron S. Anderson who conducted the tedious and meticulous dissections of the cadavers. They were enjoyable to work with and were conscientious in meeting the dissection schedule. Thanks are also extended to Elsha Russell and Dewayne Rawls for their assistance. Working with exceptional students like these adds to the enjoyment of being a professor.

It is gratifying to have professors and health-care professionals interested in the success of A Photographic Atlas for the Human Anatomy and Physiology Laboratory. There are several that were helpful in the development of this atlas. They share our enthusiasm of its value for students of anatomy and physiology. We are especially appreciative of Kyle M. Van De Graaff, M.D. at Wilford Hall in San Antonio and William B. Winborn, Ph.D. at the University of Texas Health Science Center at San Antonio for their efforts and generosity in providing the choice photomicrographs used in this atlas. The radiographs, CT scans, and MR images were made possible through the generosity of Gary M. Watts, M.D. and the Department of Radiology at Utah Valley Regional Medical Center. Kerry Peterson and Ryan L. Van De Graaff, M.D. assisted in dissections of laboratory specimens and in taking photographs. Others who aided in specimen dissections were Nathan A. Jacobson, D.O., R. Richard Rasmussen, M.D., and Sandra E. Sephton, Ph.D. We appreciate the talents of Christopher H. Creek who rendered the line art throughout the atlas. Many users and reviewers of the previous editions of this atlas provided suggestions for its improvement. We are especially appreciative of Michael J. Shively, D.V.M. for his numerous comments and helpful suggestions. We appreciate Focus Design for their help with laying out the atlas. Special thanks to Penny Dobbins of the University of Connecticut and Dr. William Ogard of the University of Alabama at Birmingham for their help in reviewing this atlas. We are indebted to Douglas Morton and the personnel at Morton Publishing Company for the opportunity, encouragement, and support to complete this project.

Anatomy is the study of body structures. An example of an anatomical study is learning about the structure of the heart—the chambers, valves, and vessels that serve the heart muscle. **Physiology** is the study of body function. An example of a physiological study is learning what causes the heart muscles to contract—the sequence of blood flow through the heart and what causes blood pressure. The anatomy (structure) and the physiology (function) of any part of the body are always related, or in other words, structure determines function.

Most of the physiological processes within the body act to maintain **homeostasis**. Simply defined, homeostasis is maintaining near consistent internal conditions within the body despite changing conditions in the external environment. For example, one area of your brain acts as a thermostat to keep your body temperature near 37°C (98.6°F). Being too warm causes you to sweat and cool the body, while being cold causes you to shiver and warm the body. Maintaining overall body homeostasis is achieved through many interacting physiological processes involving all levels of body organization, and is absolutely necessary for survival.

Structural and functional levels of organization exist in the body, and each of its parts contributes to the total organism. In the study of human anatomy and physiology, six levels of body organization are generally recognized—the molecular level, the cellular level, the tissue level, the organ level, the systems level, and the organismic level (fig. 1.1).

Cells are microscopic and are the smallest living part of all organisms. **Tissues** are layers of groups of similar cells that perform specific functions. An **organ** is an aggregate of two or more tissues integrated to perform a particular function. The **systems** of the body consist of various body organs that have similar or related functions. All the systems of the body are interrelated and function together constituting the **organism**.

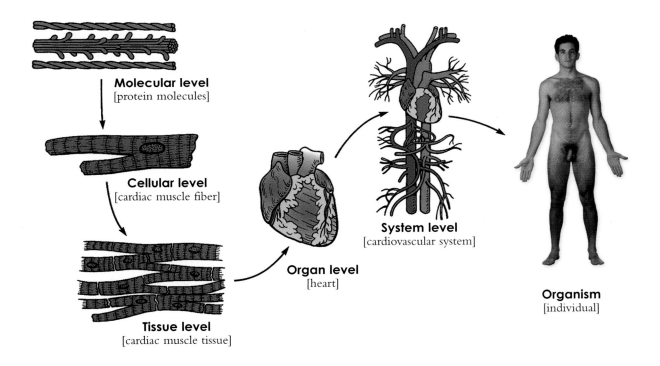

Molecular level
[protein molecules]

Cellular level
[cardiac muscle fiber]

Tissue level
[cardiac muscle tissue]

Organ level
[heart]

System level
[cardiovascular system]

Organism
[individual]

Figure 1.1
The levels of structural organization and complexity within the body.

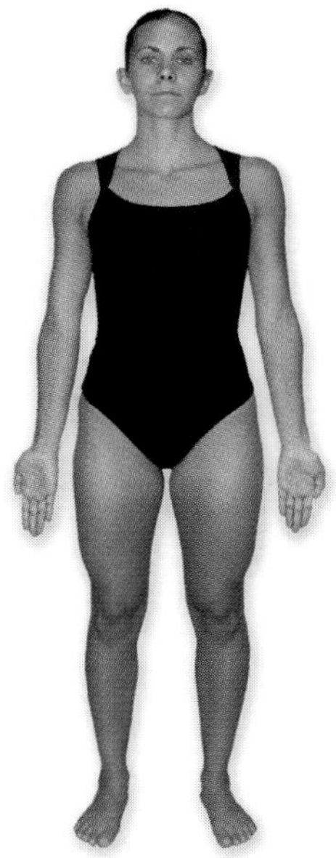

Figure 1.2
The anatomical position provides a basis of reference for describing the relationship of one body part to another. In the anatomical position, the person is standing, the feet are parallel, the eyes are directed forward, and the arms are to the sides with the palms turned forward and the fingers are pointed straight down.

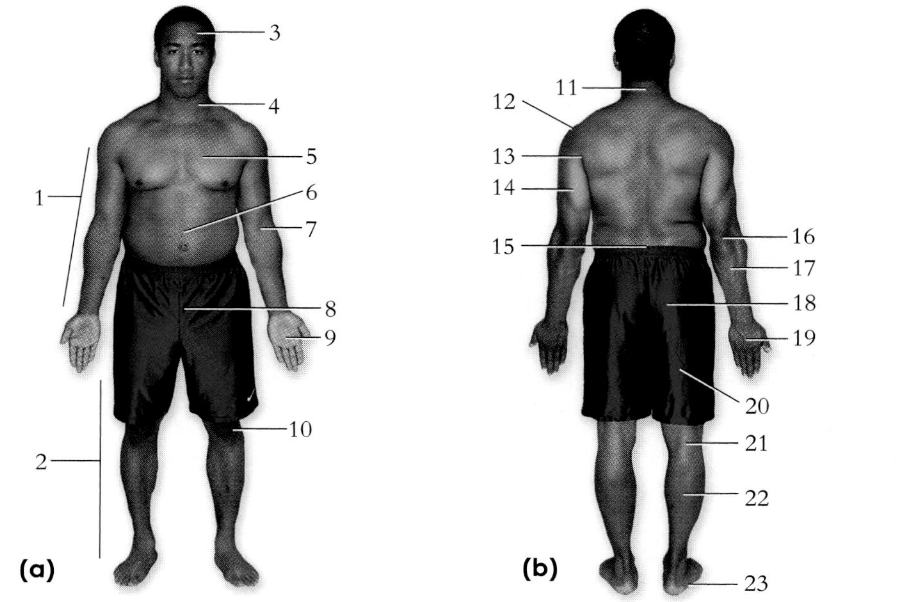

Figure 1.3
Major body parts and regions in humans (bipedal vertebrate). (a) An anterior view and (b) a posterior view.

1. Upper extremity	9. Palmar region (palm)
2. Lower extremity	10. Patellar region (patella)
3. Head	11. Cervical region
4. Neck, anterior aspect	12. Shoulder
5. Thorax (chest)	13. Axilla (armpit)
6. Abdomen	14. Brachium (upper arm)
7. Cubital fossa	15. Lumbar region
8. Pubic region	16. Elbow

17. Antebrachium (forearm)
18. Gluteal region (buttock)
19. Dorsum of hand
20. Thigh
21. Popliteal fossa
22. Calf
23. Plantar surface (sole)

Table 1.1 Directional terminology for describing human body structures.

Term	Definition	Example
Superior (cephalic, cranial)	Toward the top of the head	The neck is superior to the thorax.
Inferior (caudal)	Away from the top of the head	The pubic region is inferior to the abdomen.
Anterior (ventral)	Toward the front of the body	The eyes, nose, and mouth are on the anterior side of the body.
Posterior (dorsal)	Toward the back of the body	The spinal cord extends down the posterior side of the body.
Lateral	Toward the side of the body	The arms are on the lateral sides of the body.
Medial	Toward the median plane of the body	The heart is medial to the lungs.
Superficial (external)	Toward the surface of the body	The skin is superficial to the muscles.
Deep (internal)	Away from the surface of the body	The heart is positioned deep within the thoracic cavity.
Parietal	Reference to the body wall of the trunk (thorax and abdomen)	The parietal peritoneum is the membrane lining the abdominal cavity.
Visceral	Reference to internal organs of trunk	The stomach is covered by a thin membrane called the visceral peritoneum.

Caudal **Dorsal** **Cranial**

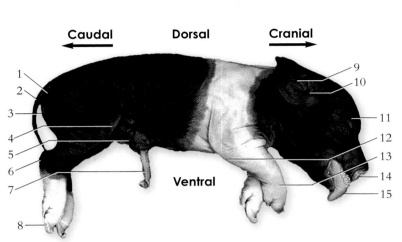

Ventral

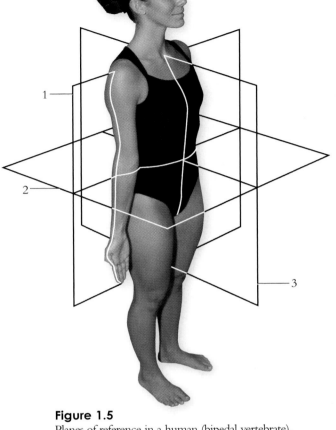

Figure 1.4
Directional terminology and superficial structures
in a fetal pig (quadrupedal vertebrate).

1. Anus
2. Tail
3. Scrotum
4. Knee
5. Teat
6. Ankle
7. Umbilical cord
8. Hoof
9. Auricle (pinna)
10. External auditory
    canal
11. Superior palpebra
    (superior eyelid)
12. Elbow
13. Wrist
14. Naris (nostril)
15. Tongue

Figure 1.5
Planes of reference in a human (bipedal vertebrate).
1. Coronal plane (frontal plane)
2. Transverse plane (cross-sectional plane)
3. Sagittal plane

Caudal **Dorsal** **Cranial**

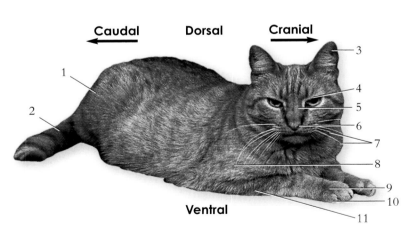

Ventral

Figure 1.6
Directional terminology and superficial
structures in a cat (quadrupedal vertebrate).

1. Thigh
2. Tail
3. Auricle (pinna)
4. Superior palpebra
   (superior eyelid)
5. Bridge of nose
6. Naris (nostril)
7. Vibrissae
8. Brachium
9. Manus (front
   foot)
10. Claw
11. Antebrachium

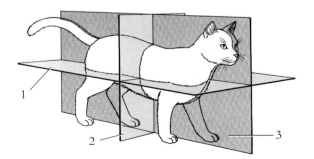

Figure 1.7
Planes of reference in a cat.
1. Coronal plane (frontal plane)
2. Transverse plane (cross-sectional plane)
3. Sagittal plane

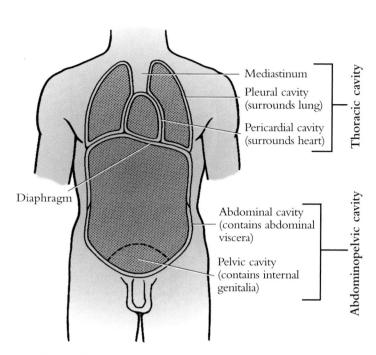

Figure 1.8
An anterior view of the body cavities of the trunk.

Mediastinum
Pleural cavity (surrounds lung)
Pericardial cavity (surrounds heart)

Thoracic cavity

Diaphragm

Abdominal cavity (contains abdominal viscera)
Pelvic cavity (contains internal genitalia)

Abdominopelvic cavity

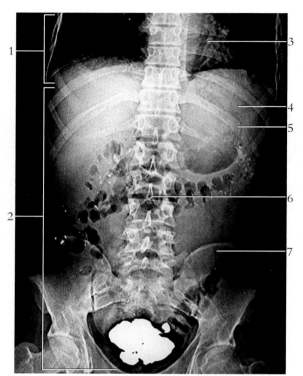

Figure 1.9
MR image of the trunk showing the body cavities and their contents.

1. Thoracic cavity
2. Abdominopelvic cavity
3. Image of heart
4. Image of diaphragm
5. Image of rib
6. Image of lumbar vertebra
7. Image of Ilium

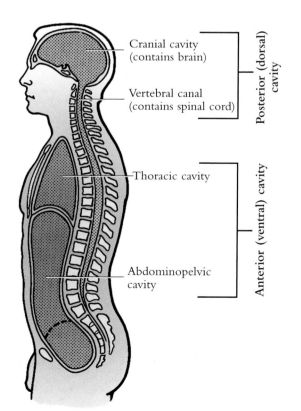

Figure 1.10
A midsagittal view of the body cavities.

Cranial cavity (contains brain)
Vertebral canal (contains spinal cord)

Posterior (dorsal) cavity

Thoracic cavity

Abdominopelvic cavity

Anterior (ventral) cavity

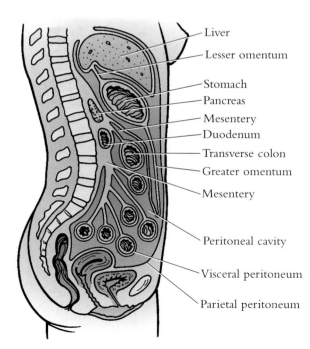

Figure 1.11
A midsagittal view of the organs of the abdominopelvic cavity and their supporting membranes.

Liver
Lesser omentum
Stomach
Pancreas
Mesentery
Duodenum
Transverse colon
Greater omentum
Mesentery
Peritoneal cavity
Visceral peritoneum
Parietal peritoneum

Figure 1.12

Human male.
(a) Anterior view
(b) Posterior view
 1. Facial region
 2. Cranial region
 3. Posterior neck
 4. Anterior neck
 5. Shoulder
 6. Thorax
 7. Nipple
 8. Brachium
 9. Elbow
10. Cubital fossa
11. Abdomen
12. Umbilicus (navel)
13. Antebrachium
14. Wrist
15. Hand
16. Natal (gluteal) cleft
17. Fold of buttock (gluteal fold)
18. External genitalia
19. Thigh
20. Patella
21. Popliteal fossa
22. Leg
23. Ankle
24. Foot

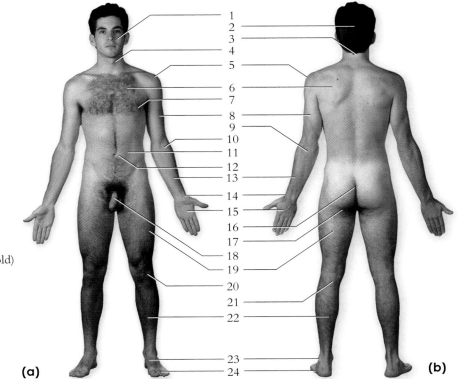

(a)　　　　**(b)**

Figure 1.13

Human female.
(a) Anterior view
(b) Posterior view
 1. Facial region
 2. Cranial region
 3. Posterior neck
 4. Anterior neck
 5. Shoulder
 6. Thorax
 7. Breast
 8. Nipple
 9. Brachium
10. Cubital fossa
11. Elbow
12. Abdomen
13. Antebrachium
14. Iliac crest
15. Umbilicus (navel)
16. Wrist
17. Hand
18. Natal (gluteal) cleft
19. Fold of buttock (gluteal fold)
20. Mons pubis
21. Thigh
22. Popliteal fossa
23. Patella
24. Leg
25. Ankle
26. Foot

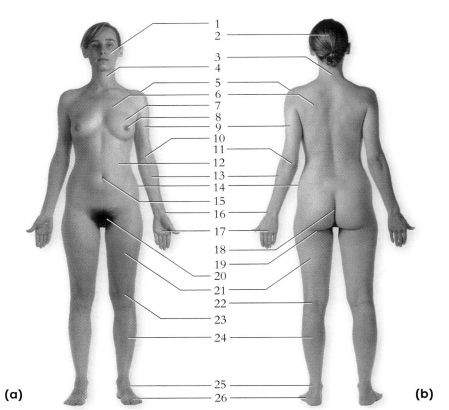

(a)　　　　**(b)**

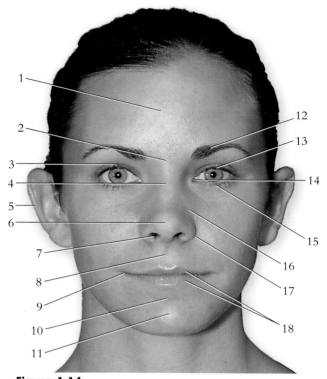

Figure 1.14
An anterior view of the facial region.

1. Forehead
2. Root of nose (glabella)
3. Superior palpebral sulcus
4. Bridge of nose
5. Auricle (pinna)
6. Apex of nose
7. Nostril
8. Philtrum
9. Corner of mouth
10. Mentolabial sulcus
11. Chin (mentalis)
12. Eyebrow
13. Eyelashes of upper eyelid
14. Lacrimal caruncle
15. Eyelashes of lower eyelid
16. Nasofacial angle
17. Alar nasal sulcus
18. Lips

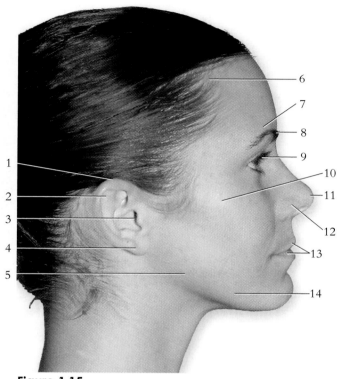

Figure 1.15
A lateral view of the facial region.

1. Helix of auricle
2. Antihelix
3. External auditory canal
4. Earlobe
5. Angle of mandible
6. Hair line
7. Superciliary ridge
8. Eyebrow
9. Eyelashes
10. Zygomatic arch
11. Apex of nose
12. Ala nasi
13. Lips
14. Body of mandible

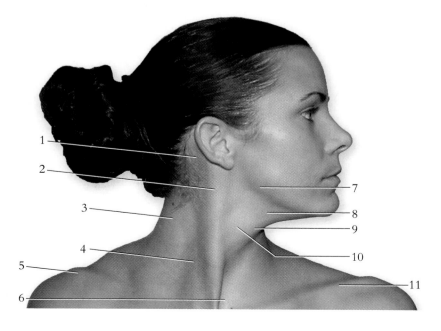

Figure 1.16
An anterolateral view of the neck.
(m. = muscle)

1. Mastoid process
2. Sternocleidomastoid m.
3. Trapezius m.
4. Posterior triangle of neck
5. Acromion of scapula
6. Jugular notch
7. Angle of mandible
8. Hyoid bone
9. Thyroid cartilage of larynx
10. Anterior triangle of neck
11. Clavicle

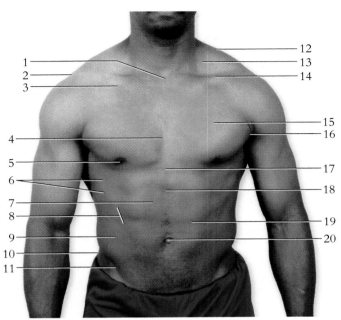

Figure 1.17
An anterior view of the thorax and abdomen.

1. Jugular notch
2. Acromion of scapula
3. Clavipectoral triangle
4. Sternum
5. Nipple
6. Serratus anterior m.
7. Rectus abdominis m.
8. Linea semilunaris
9. External abdominal oblique m.
10. Iliac crest
11. Inguinal ligament
12. Trapezius m.
13. Supraclavicular fossa
14. Clavicle
15. Pectoralis major m.
16. Anterior axillary fold
17. Xiphoid process
18. Linea alba
19. Tendinous inscription
20. Umbilicus

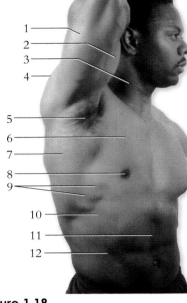

Figure 1.18
An anterolateral view of the thorax, abdomen, and axilla.

1. Triceps brachii m.
2. Biceps brachii m.
3. Sternocleidomastoid m.
4. Deltoid m.
5. Axilla
6. Pectoralis major m.
7. Latissimus dorsi m.
8. Nipple
9. Serratus anterior m.
10. Intercostal m.
11. Linea alba
12. External abdominal oblique m.

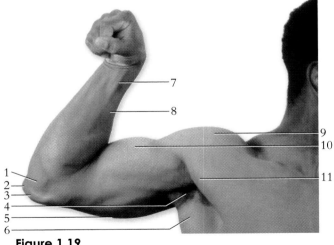

Figure 1.19
Right shoulder, axilla, and upper extremity.

1. Medial epicondyle of humerus
2. Olecranon of ulna
3. Ulnar Sulcus
4. Axilla
5. Triceps brachii m.
6. Latissimus dorsi m. (posterior axillary fold)
7. Tendon of flexor carpi radialis longus m.
8. Brachioradialis m.
9. Deltoid m.
10. Biceps brachii m.
11. Pectoralis major m. (anterior axillary fold)

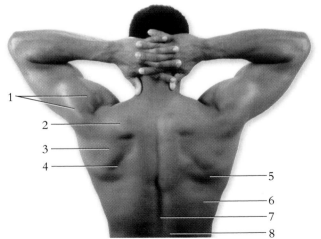

Figure 1.20
A posterior view of the thorax.

1. Deltoid m.
2. Trapezius m.
3. Infraspinatus m.
4. Triangle of auscultation
5. Inferior angle of scapula
6. Latissimus dorsi m.
7. Median furrow over vertebral column
8. Erector spinae m.

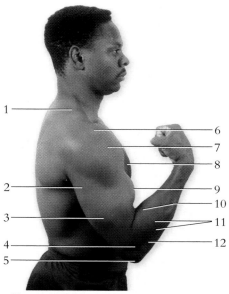

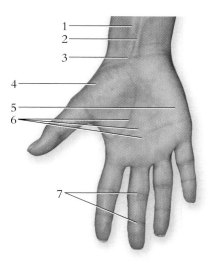

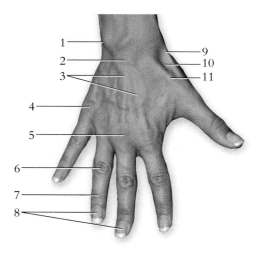

Figure 1.21
A lateral view of the right shoulder and upper extremity.
(m. = muscle, mm. = muscles)
1. Trapezius m.
2. Long head of triceps brachii m.
3. Lateral head of triceps brachii m.
4. Lateral epicondyle of humerus
5. Olecranon of ulna
6. Acromion of scapula
7. Deltoid m.
8. Pectoralis major m.
9. Biceps brachii m.
10. Brachioradialis m.
11. Extensor carpi radialis longus and brevis mm.
12. Extensor digitorum m.

Figure 1.22
An anterior view of the right hand.
1. Tendon of flexor carpi radialis m.
2. Tendon of palmaris longus m.
3. Flexion crease on wrist
4. Thenar eminence
5. Hypothenar eminence
6. Flexion creases on palm of hand
7. Flexion creases on third digit

Figure 1.23
A posterior view of the right hand.
1. Styloid process of ulna
2. Position of extensor retinaculum
3. Tendons of extensor digitorum m.
4. Tendon of extensor digiti minimi m.
5. Metacarpophalangeal joint
6. Proximal interphalangeal joint
7. Distal interphalangeal joint
8. Nails
9. Tendon of extensor pollicis brevis m.
10. Anatomical snuffbox
11. Tendon of extensor pollicis longus m.

Figure 1.24
An anterior view of the right upper extremity.
1. Cephalic vein
2. Biceps brachii m.
3. Cubital fossa
4. Brachioradialis m.
5. Cephalic vein
6. Site for palpation of radial artery
7. Tendon of flexor carpi radialis m.
8. Tendon of palmaris longus m.
9. Thenar eminence
10. Metacarpophalangeal joint of thumb
11. Site of palpation of brachial artery
12. Basilic vein
13. Median cubital vein
14. Ulnar vein
15. Median antebrachial vein
16. Tendon of superficial digital flexor m.
17. Styloid process of ulna
18. Hypothenar eminence

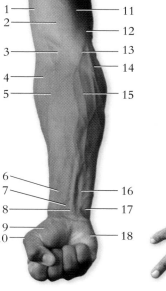

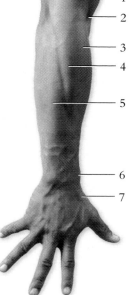

Figure 1.25
A posterior view of the right upper extremity.
1. Biceps brachii m.
2. Cubital fossa
3. Brachioradialis m.
4. Extensor carpi radialis longus m.
5. Extensor carpi ulnaris m.
6. Styloid process of radius
7. Tendon of extensor pollicus longus m.

Figure 1.26

An anterior view of
the right thigh.
1. Site of femoral triangle
2. Quadriceps femoris
   group of muscles
3. Adductor group of
   muscles
4. Rectus femoris m.
5. Vastus lateralis m.
6. Sartorius m.
7. Vastus medialis m.
8. Tendon of quadriceps
   femoris m.
9. Patella
10. Patellar ligament

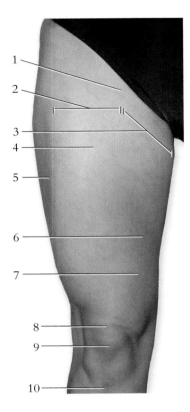

Figure 1.27

A medial view of
the right thigh.
1. Adductor magnus m.
2. Gracilis m.
3. Rectus femoris m.
4. Sartorius m.
5. Vastus medialis m.
6. Semimembranosus m.
7. Semitendinosus m.
8. Patella

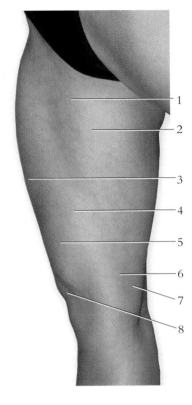

Figure 1.28

A posterior view of
the right thigh.
1. Gluteus maximus m.
2. Fold of buttock
   (gluteal fold)
3. Hamstring group
   of muscles
4. Vastus lateralis m.
5. Long head of biceps
   femoris m.
6. Semitendinosus m.
7. Gracilis m.
8. Popliteal fossa
9. Lateral head of
   gastrocnemius m.

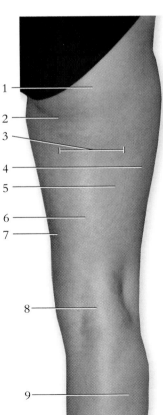

Figure 1.29

A lateral view of
the right thigh.
1. Gluteus maximus m.
2. Tensor fasciae latae m.
3. Vastus lateralis m.
4. Rectus femoris m.
5. Biceps femoris m.
6. Iliotibial tract
7. Patella
8. Lateral epicondyle
   of femur
9. Popliteal fossa

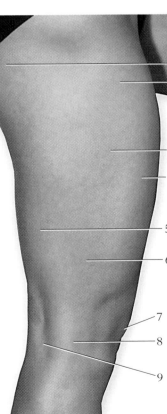

Figure 1.30

An anterior view of
the right leg and foot.
1. Patella
2. Patellar ligament
3. Tibialis anterior m.
4. Lateral malleolus of fibula
5. Medial malleolus of tibia
6. Site for palpation of
   dorsal pedis artery
7. Tendons of extensor
   digitorum longus m.
8. Tendon of extensor
   hallucis longus m.

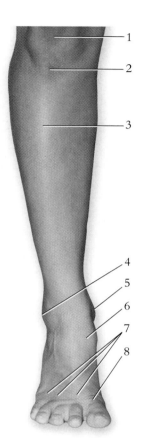

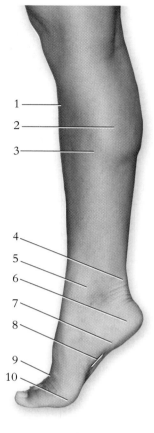

Figure 1.31

A medial view of
the right leg and foot.
1. Tibia
2. Medial head of
   gastrocnemius m.
3. Soleus m.
4. Tendo calcaneus
5. Medial malleolus of tibia
6. Calcaneus
7. Abductor hallucis m.
8. Longitudinal arch
9. Tendon of extensor
   hallucis longus m.
10. Head of first
    metatarsal bone

Figure 1.32

A posterior view of
the right leg and foot.
1. Popliteal fossa
2. Lateral head of
   gastrocnemius m.
3. Medial head of
   gastrocnemius m.
4. Soleus m.
5. Peroneus longus m.
6. Tendo calcaneus
7. Peroneus brevis m.
8. Medial malleolus of tibia
9. Lateral malleolus
   of fibula
10. Calcaneus
11. Abductor digiti
    minimi m.
12. Plantar surface of foot

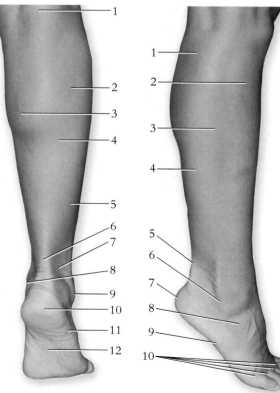

Figure 1.33

A lateral view of
the right leg and foot.
1. Lateral head of
   gastrocnemius m.
2. Tibialis anterior m.
3. Peroneus longus m.
4. Soleus m.
5. Tendo calcaneus
6. Lateral malleolus
   of fibula
7. Calcaneus
8. Extensor digitorum
   brevis m.
9. Lateral surface of foot
10. Tendons of extensor
    digitorum longus m.

Cells are the basic structural and functional units of organization within the body. Although diverse, human cells have structural similarities including a **nucleus** containing a nucleolus, various **organelles** suspended in **cytoplasm**, and an enclosing cell (plasma) **membrane** (fig. 2.1).

The **nucleus** is the large spheroid body within a cell that contains the **chromatin**, **nucleolus**, and **nucleoplasm**—the genetic material of the cell. The nucleus is enclosed by a double membrane called the **nuclear membrane**, or **nuclear envelope**. The **nucleolus** is a dense, nonmembranous body composed of protein and RNA molecules. The chromatin consists of fibers of protein and DNA molecules. Prior to cellular division, the chromatin shortens and coils into rod-shaped **chromosomes**. Chromosomes consists of DNA and proteins called histones.

The **cytoplasm** of a cell is the medium of support between the nuclear membrane and the cell membrane. **Organelles** are minute membrane-bound structures within the cytoplasm of a cell that are concerned with specific functions. The cellular functions carried out by the organelles are referred to as **cellular metabolism**. The principal organelles and their functions are listed in table 2.1. In order for cells to remain alive, metabolize, and maintain homeostasis, certain requirements must be met. These include having access to nutrients and oxygen, being able to eliminate wastes, and being maintained in a constant, protective environment.

The **cell membrane** is composed of phospholipid and protein molecules, which gives form to a cell and controls the passage of material into and out of a cell. More specifically, the proteins in the cell membrane provide: 1) structural support; 2) a mechanism of molecule transport across the membrane; 3) enzymatic control of chemical reactions; 4) receptors for hormones and other regulatory molecules; and 5) cellular markers (antigens), which identify the blood and tissue type. The carbohydrate molecules: 1) repel negative objects due to their negative charge; 2) act as receptors for hormones and other regulatory molecules; 3) form specific cell markers which enable like cells to attach and aggregate into tissues; and 4) enter into immune reactions.

The permeability of the cell membrane is a function of: 1) size of molecules; 2) solubility in lipids; 3) ionic charge of molecules; and 4) the presence of carrier molecules.

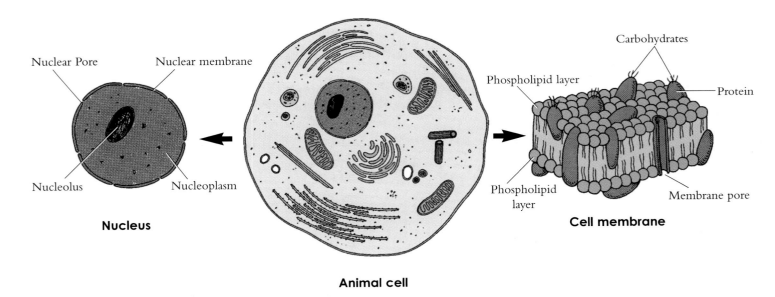

Nucleus

Nuclear Pore   Nuclear membrane

Nucleolus   Nucleoplasm

Animal cell

Cell membrane

Carbohydrates

Phospholipid layer

Protein

Phospholipid layer

Membrane pore

Figure 2.1
A cell and it's nucleus and cell (plasma) membrane.

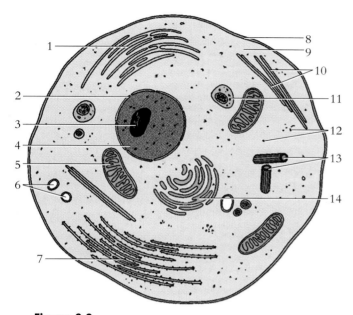

Figure 2.2
A typical animal cell.
1. Smooth endoplasmic reticulum
2. Z
3. Nucleolus
4. Nucleoplasm
5. Mitochondrion
6. Vesicles
7. Rough endoplasmic reticulum
8. Cell membrane
9. Cytoplasm
10. Microtubules
11. Lysosome
12. Ribosomes
13. Centrioles
14. Golgi complex

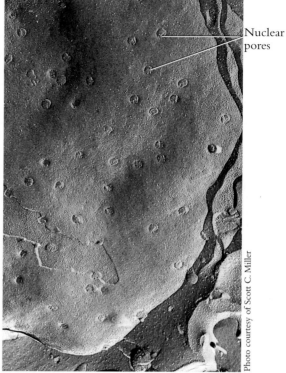

Nuclear pores

Photo courtesy of Scott C. Miller

Figure 2.3   1000X
An electron micrograph of a freeze fractured
nuclear envelope showing the nuclear pores.

Table 2.1 Structure and function of cellular components.

Component	Structure	Function
Cell (plasma) membrane	Composed of protein and phospholipid molecules	Provides form to cell and controls passage of materials into and out of cell
Cytoplasm	Fluid to jelly-like substance	Suspends organelles and provides a matrix in which chemical reactions occur
Endoplasmic reticulum	Interconnecting hollow channels	Provides supporting framework of cell; facilitates cell transport
Ribosomes	Granules of nucleic acid	Synthesize proteins
Mitochondria	Double-layered sacs with cristae	Production of ATP
Golgi complex	Flattened sacs with vacuoles	Synthesize carbohydrates and packages molecules for secretion
Lysosomes	Membrane-surrounded sacs of enzymes	Digest foreign molecules and worn cells
Centrosome	Mass of two rodlike centrioles	Organizes spindle fibers and assists mitosis
Vacuoles	Membranous sacs	Store and excrete substances within the cytoplasm
Fibrils and microfibrils	Protein strands	Support cytoplasm and transport materials
Cilia and flagella	Cytoplasmic extensions from cell	Movement of particles along cell surface or move cell
Nucleus	Nuclear membrane, nucleolus, and chromatin (DNA)	Directs cell activity; forms ribosomes

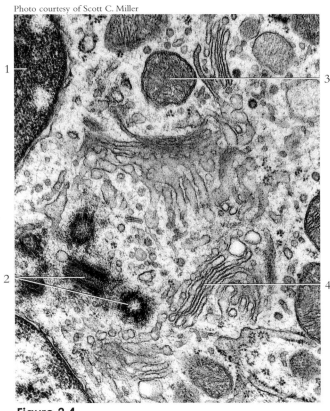

Photo courtesy of Scott C. Miller

Figure 2.4
An electron micrograph of various organelles.
1. Nucleus   3. Mitochondrion
2. Centrioles   4. Golgi complex

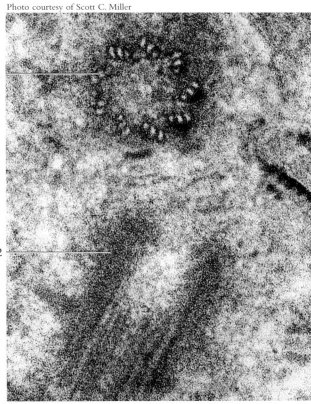

Photo courtesy of Scott C. Miller

Figure 2.5
An electron micrograph of centrioles. The centrioles
are positioned at right angles to one another.
1. Centriole (shown in   2. Centriole (shown in
   cross section)   longitudinal section)

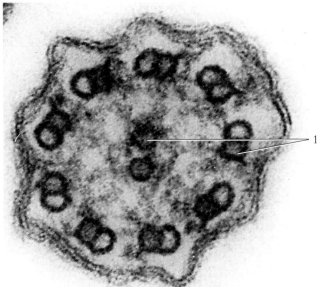

Photo courtesy of Scott C. Miller

Figure 2.6
An electron micrograph of cilia (cross section) showing the
characteristic "9 + 2" arrangement of microtubules in the
cross sections.
1. Microtubules

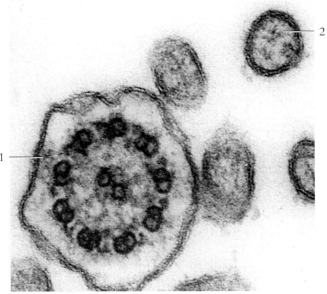

Photo courtesy of Scott C. Miller

Figure 2.7
An electron micrograph showing the difference between
a microvillus and a cilium.
1. Microvillus   2. Cilium

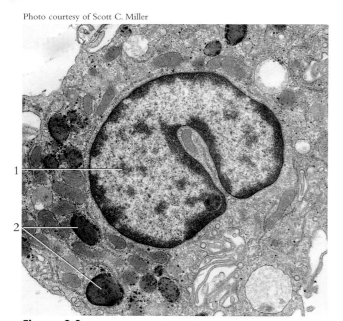

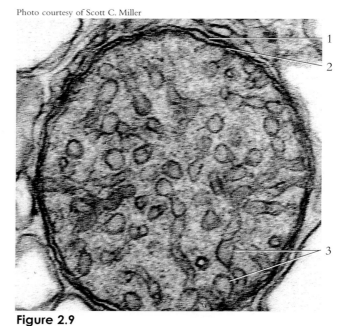

Figure 2.8
An electron micrograph of lysosomes.
1. Nucleus
2. Lysosomes

Figure 2.9
An electron micrograph of a mitochondrion.
1. Outer membrane
2. Crista
3. Inner membranes

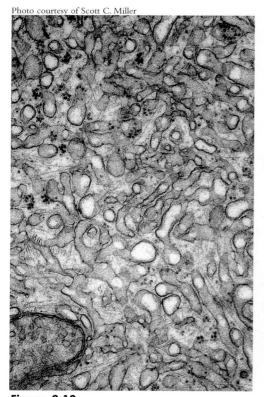

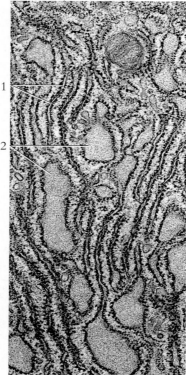

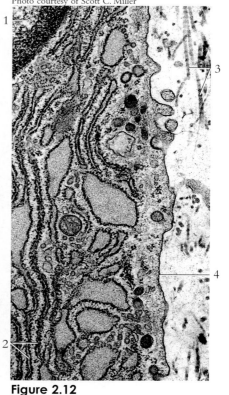

Figure 2.10
An electron micrograph of smooth endoplasmic reticulum from the testis.

Figure 2.11
An electron micrograph of rough endoplasmic reticulum.
1. Ribosomes
2. Cisternae

Figure 2.12
Rough endoplasmic reticulum secreting collagenous filaments to the outside of the cell.
1. Nucleus
2. Rough endoplasmic reticulum
3. Collagenous filaments
4. Cell membrane

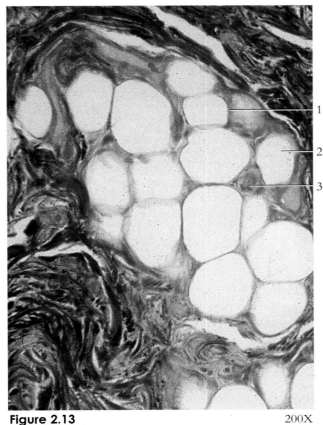

Figure 2.13   200X
Adipocytes (fat cells) in adipose tissue.
  1. Cell membrane of adipocyte
  2. Lipid-filled vacuole of adipocyte
  3. Nucleus

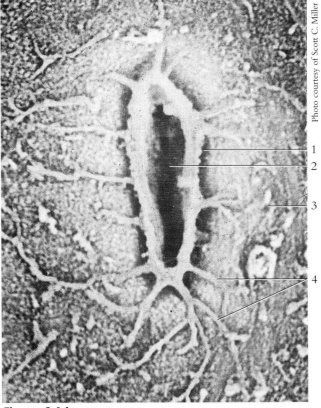

Photo courtesy of Scott C. Miller

Figure 2.14
An electron micrograph of an osteocyte (bone cell)in cortical
bone matrix.
  1. Lacuna   3. Bone matrix
  2. Osteocyte   4. Canaliculi

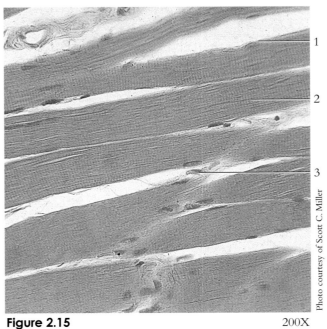

Photo courtesy of Scott C. Miller

Figure 2.15   200X
Skeletal muscle cells (fibers).
  1. Sarcolemma   3. Nucleus
  2. Striations

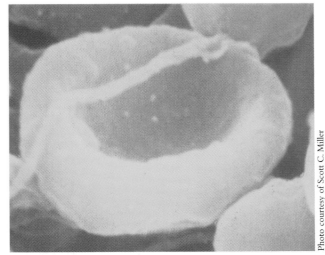

Photo courtesy of Scott C. Miller

Figure 2.16
An electron micrograph of an erythrocyte (red blood cell).

Photo courtesy of Scott C. Miller

Figure 2.17
An electron micrograph of a skeletal muscle myofibril, showing the striations.

1. Mitochondria   4. I band   7. H band
2. Z line   5. T-tubule   8. Sacromere
3. A band   6. Sarcoplasmic reticulum

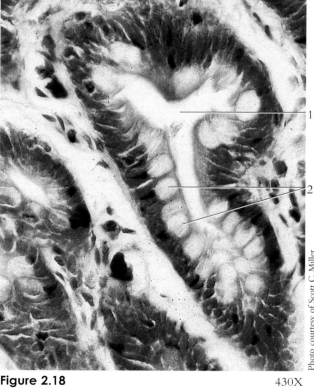

Photo courtesy of Scott C. Miller

Figure 2.18   430X
Goblet cells within an intestinal gland
(crypt of Lieberkühn) of small intestine.
1. Lumen of gland
2. Goblet cells

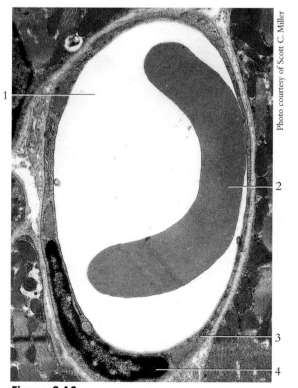

Photo courtesy of Scott C. Miller

Figure 2.19
An electron micrograph of a capillary
containing an erythrocyte.
1. Lumen of capillary   3. Endothelial cell
2. Erythrocyte   4. Nucleus of endothelial cell

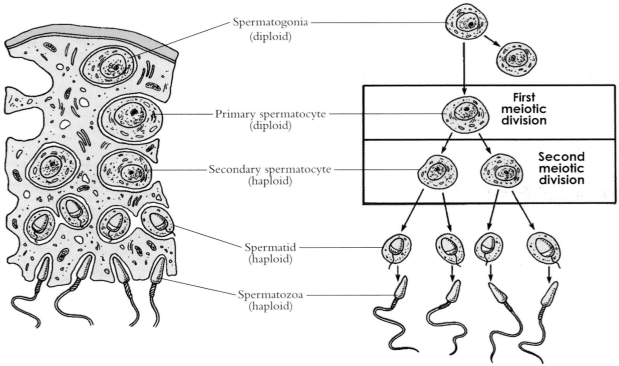

Figure 2.20
Spermatogenesis is the production of male gametes, or spermatozoa, through the process of meiosis.

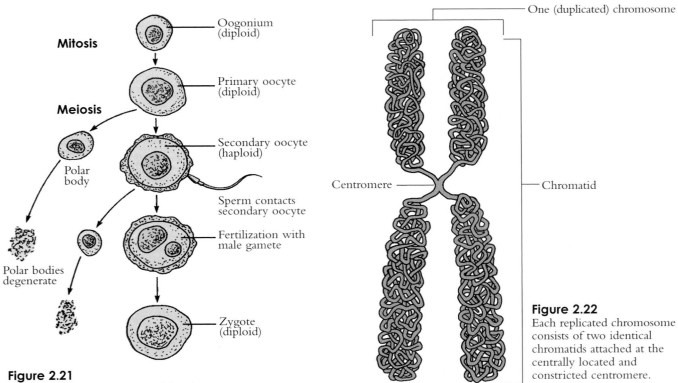

Figure 2.21
Oogenesis is the production of female sex gametes, or ova, through the process of meiosis.

Figure 2.22
Each replicated chromosome consists of two identical chromatids attached at the centrally located and constricted centromere.

Figure 2.23
Stages of mitosis.

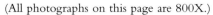

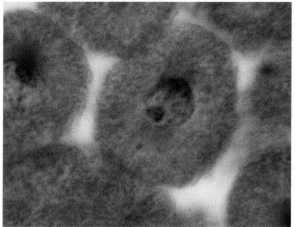

Prophase
Each chromosome consists of two chromatids joined by a centromere. Spindle fibers extend from each centriole.

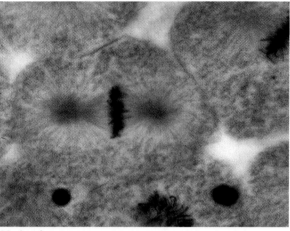

Metaphase
The chromosomes are positioned at the equator. The spindle fibers from each centriole attach to the centromeres.

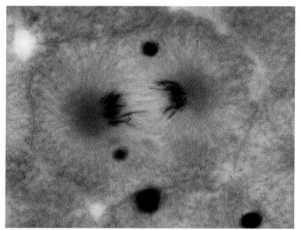

Anaphase
The centromeres split, and the sister chromatids separate as each is pulled to an opposite pole.

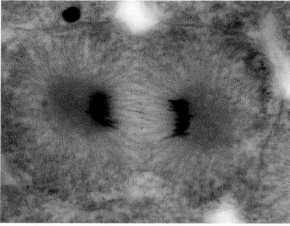

Telophase
The chromosomes lengthen and become less distinct. The cell membrane forms between the forming daughter cells.

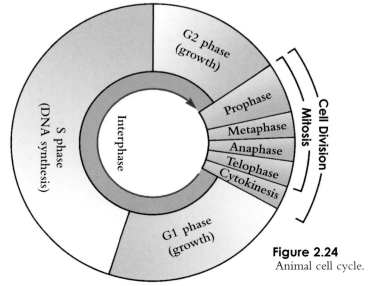

Figure 2.24
Animal cell cycle.

Histology 3

While it is true that cells comprise the basic structural and functional units of the body, the cells in a multicellular organism, such as a human, are so specialized that they do not function independently. **Tissues** are aggregations of similar cells that perform specific functions. **Histology** is the science concerned with the study of cells. Both **cytology**, the study of cells, and histology are actually microscopic anatomy. Although cytologists and histologists utilize many different techniques to study cells and tissues, basically only two kinds of microscopes are used the view the prepared specimens. **Light microscopy** is used for the general observation of cellular and tissue structure, and **electron microscopy** permits observation of the fine details of the specimens.

In electron microscopy, a beam of electrons is passed through an object in a procedure called **transmission electron microscopy (TEM)**, or the beam is reflected off the surface of an object in a procedure called **scanning electron microscopy (SEM)**. In both cases, the electron beam is magnified with electromagnets. The depth of focus of SEM is much greater than it is with TEM, producing a clear three-dimensional image of cellular or tissue structure. The magnification ability of SEM, however, is not as great as that of TEM.

The tissues of the body are classified into four principal types, determined by structure and function: 1) **epithelial tissues** cover body and organ surfaces, line body and luminal (hollow portion of body tubes) cavities, and form various glands; 2) **connective tissues** bind, support, and protect body parts; 3) **muscle tissues** contract to produce movements; and 4) **nervous tissues** initiate and transmit nerve impulses from one body part to another.

Epithelial tissues are classified by the number of layers of cells and the shapes of the cells along the exposed surfaces. A **simple epithelial tissue** is made up of a single layer of cells. A **stratified epithelial tissue** is made up of layers of cells. The basic shapes of the exposed cells are: **squamous**, or flattened; **cuboidal**, or cube-shaped; and **columnar**, or elongated.

Connective tissues are classified according to the characteristics of the **matrix**, or binding material between the similar cells. The classification of connective tissues is not exact, but the following is a commonly accepted scheme of classification:

A. **Embryonic connective tissue**

B. **Connective tissue proper**
   1. Loose (areolar) connective tissue
   2. Dense regular connective tissue
   3. Dense irregular connective tissue
   4. Elastic connective tissue
   5. Reticular connective tissue
   6. Adipose tissue

C. **Cartilage**
   1. Hyaline cartilage
   2. Fibrocartilage cartilage
   3. Elastic cartilage

D. **Bone tissue**

E. **Blood (vascular tissue)**

Muscle tissues are responsible for the movement of materials through the body, the movement of one part of the body with respect to another, and for locomotion. The three kinds of muscle tissue are **smooth**, **cardiac**, and **skeletal**. The fibers in all three kinds are adapted to contract in response to stimuli.

Nervous tissues are composed of **neurons**, which respond to stimuli and conduct action potentials (nerve impulses) to and from all body organs, and **neuroglia**, which functionally support and physically bind neurons.

Figure 3.1   300X
Simple squamous epithelium.
  1. Single layer of flattened cells

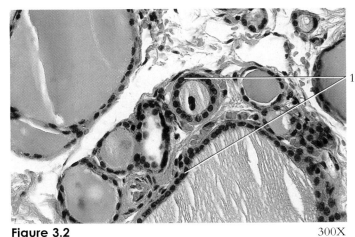

Figure 3.2   300X
Simple cuboidal epithelium.
  1. Single layer of cells with round nuclei

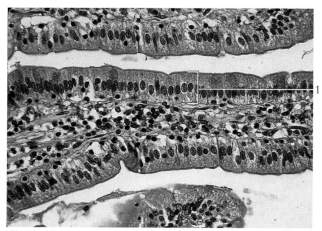

Figure 3.3　　　　　　　　　　300X
Simple columnar epithelium.
 1.　Single layer of cells with oval nuclei

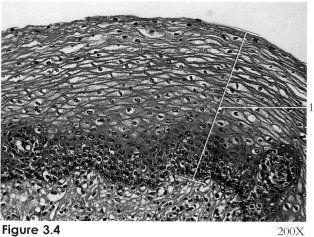

Figure 3.4　　　　　　　　　　200X
Stratified squamous epithelium.
 1.　Multiple layers of cells, which are flattened at the upper layer

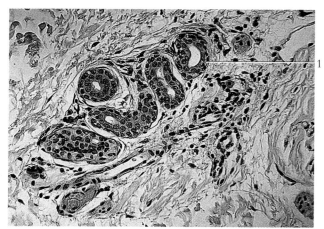

Figure 3.5　　　　　　　　　　200X
Stratified cuboidal epithelium.
 1.　Two layers of cells with round nuclei

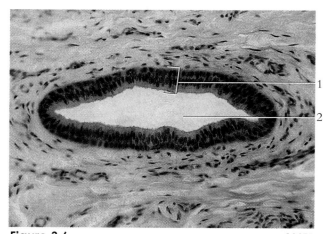

Figure 3.6　　　　　　　　　　200X
Stratified columnar epithelium.
 1.　Two layers of cells with oval nuclei
 2.　Lumen

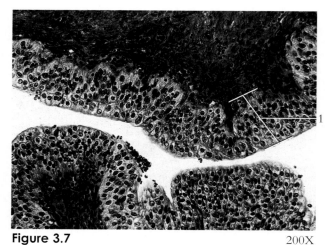

Figure 3.7　　　　　　　　　　200X
Stratified columnar epithelium.
 1.　Cells are balloon-like at surface

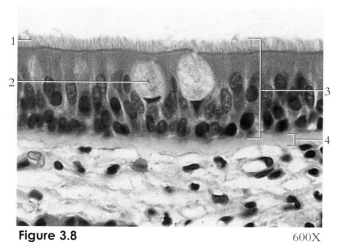

Figure 3.8　　　　　　　　　　600X
Pseudostratified columnar epithelium.
 1.　Cilia
 2.　Goblet cell
 3.　Pseudostratified columnar epithelium
 4.　Basement membrane

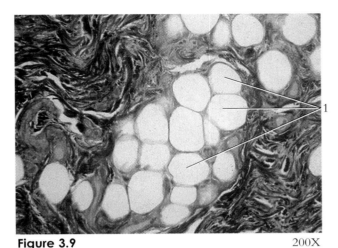

Figure 3.9   200X
Adipose connective tissue.
1. Adipocytes (adipose cells)

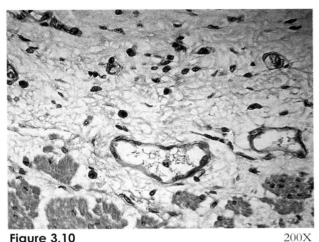

Figure 3.10   200X
Loose connective tissue.

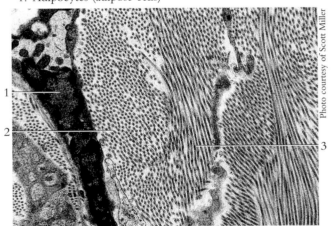

Photo courtesy of Scott Miller

Figure 3.11
An electron micrograph of collagenous fibers in loose
connective tissue.
1. Fibroblast
2. Cross section of bundle of collagenous fibers
3. Longitudinal section of bundle of collagenous fibers

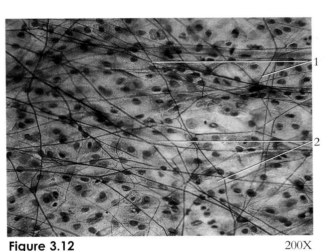

Figure 3.12   200X
Loose connective tissue stained for fibers.
1. Elastic fibers (black)
2. Collagen fibers (pink)

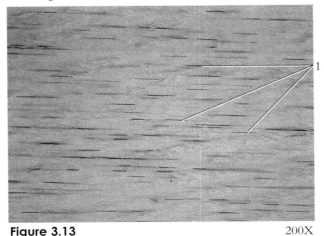

Figure 3.13   200X
Dense regular connective tissue.
1. Nuclei of fibroblasts arranged in parallel rows

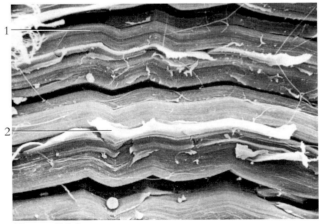

Figure 3.14
Electron micrograph of dense regular connective tissue.
1. Collagenous fiber
2. Fibroblast

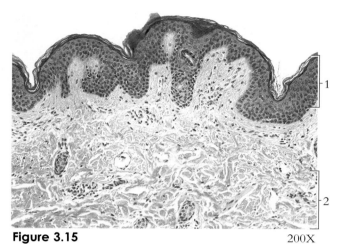

Figure 3.15   200X
Dense irregular connective tissue.
  1. Epidermis
  2. Dense irregular connective tissue (reticular layer of dermis)

Figure 3.16
Electron micrograph of dense irregular connective tissue.
  1. Collagenous fibers

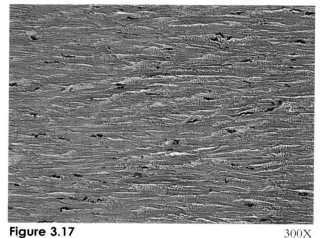

Figure 3.17   300X
Dense irregular connective tissue.

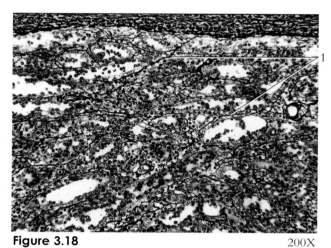

Figure 3.18   200X
Reticular connective tissue.
  1. Reticular fibers

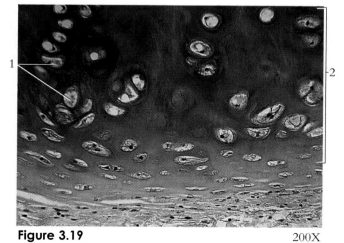

Figure 3.19   200X
Hyaline cartilage.
  1. Chondrocytes
  2. Hyaline cartilage

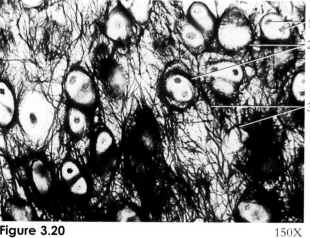

Figure 3.20   150X
Elastic cartilage.
  1. Chondrocytes
  2. Lacunae
  3. Elastic fibers

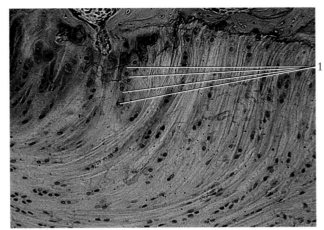

Figure 3.21 150X
Fibrocartilage.
1. Chondrocytes arranged in a row

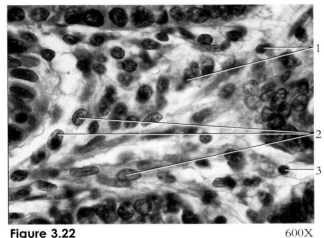

Figure 3.22 600X
Cells of connective tissue.
1. Eosinophils   2. Fibroblasts   3. Lymphocyte

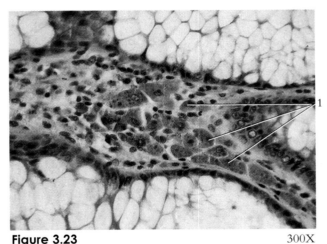

Figure 3.23 300X
Cells of connective tissue.
1. Macrophages

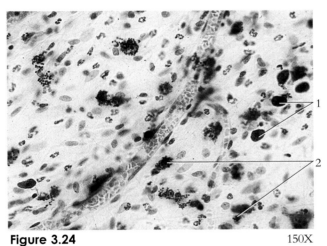

Figure 3.24 150X
Cells of connective tissue, special preparation.
1. Mast cells   2. Macrophages

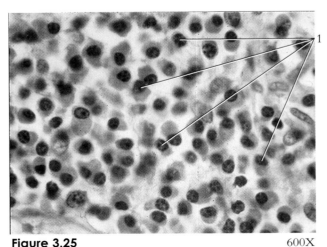

Figure 3.25 600X
Cells of connective tissue.
1. Plasma cells

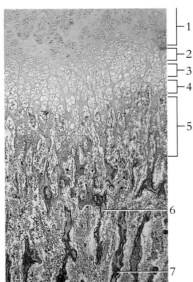

Figure 3.26
Endochondral bone formation.
1. Zone of reserve cartilage
2. Zone of proliferation
3. Zone of hypertrophy
4. Zone of calcification
5. Zone of resorption
6. Calcified cartilage (blue)
7. Spicule of bone (red)

40X

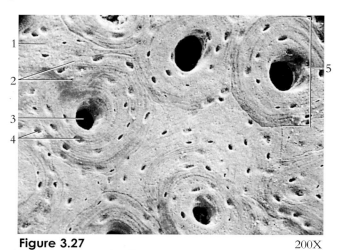

Figure 3.27   200X
Electron micrograph of bone tissue.
1. Interstitial lamellae   4. Lacunae
2. Lamellae   5. Osteon (haversian
3. Central canal   system)
   (haversian canal)

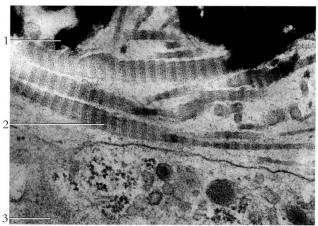

Figure 3.28
SEM photomicrograph of bone tissue formation.
1. Bone mineral (calcium salts stain black)
2. Collagenous filament (distinct banding pattern)
3. Collagen secreting osteoblasts

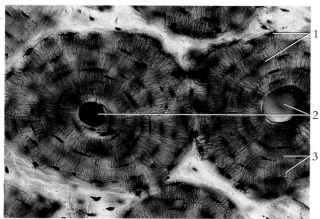

Figure 3.29   200X
Cross section of two osteons.
1. Lacunae   3. Lamellae
2. Central (haversian) canals

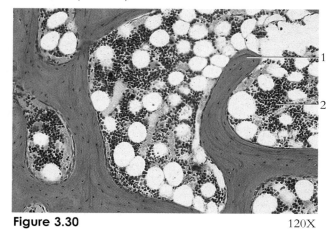

Figure 3.30   120X
Spongy (cancellous) bone.
1. Spicule of bone
2. Marrow (includes adipose cells)

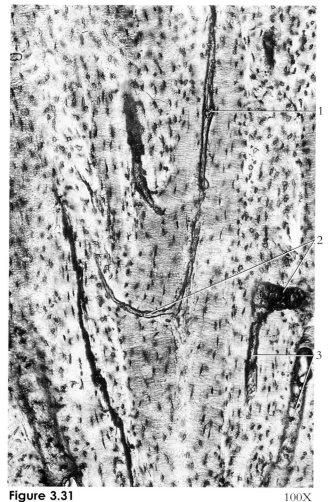

Figure 3.31   100X
Longitudinal section of osteons.
1. Central (haversian) canal
2. Perforating (Volkman's) canals
3. Central (haversian) canals

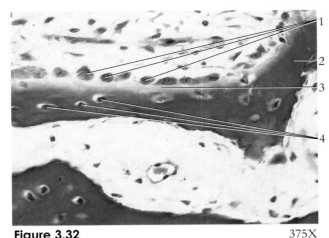

Figure 3.32   375X
Osteoblasts.
1. Osteoblasts   3. Osteoid
2. Bone   4. Osteocytes

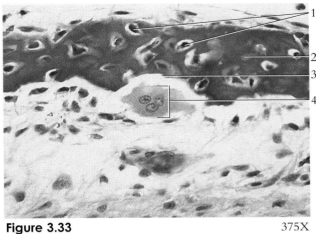

Figure 3.33   375X
Osteoclast.
1. Osteocytes   4. Osteoclast in
2. Bone   Howship's lacuna
3. Howship's lacuna

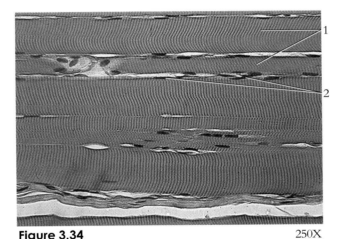

Figure 3.34   250X
Longitudinal section of skeletal muscle tissue.
1. Skeletal muscle cells, note striations
2. Multiple nuclei in periphery of cell

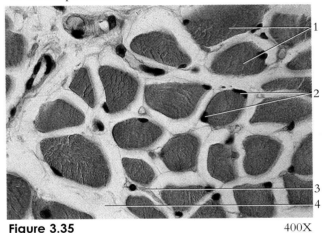

Figure 3.35   400X
Cross section of skeletal muscle tissue.
1. Skeletal muscle cells
2. Nuclei in periphery of cell
3. Endomysium (surrounds cells)
4. Perimysium (surrounds bundles of cells)

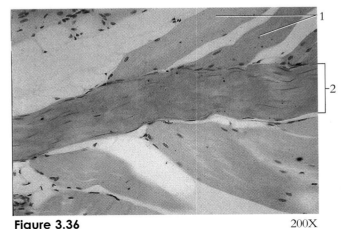

Figure 3.36   200X
Attachment of skeletal muscle to tendon.
1. Skeletal muscle
2. Dense regular connective tissue (tendon)

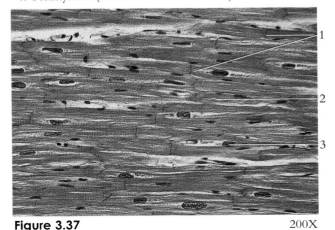

Figure 3.37   200X
Cardiac muscle tissue.
1. Intercalated discs
2. Light-staining perinuclear sarcoplasm
3. Nucleus in center of cell

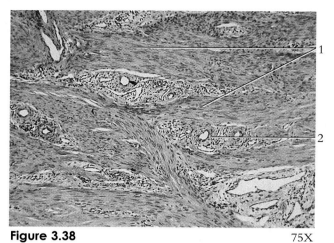

Figure 3.38 75X
Smooth muscle tissue.
  1. Smooth muscle
  2. Blood vessel

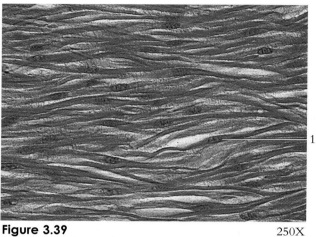

Figure 3.39 250X
Partially teased smooth muscle tissue.
  1. Nucleus of individual cell

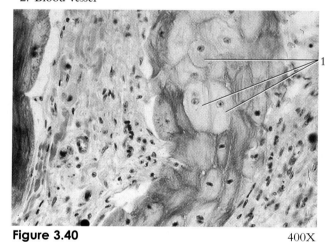

Figure 3.40 400X
Conduction myofibers (Purkinje fibers).
  1. Conduction myofibers

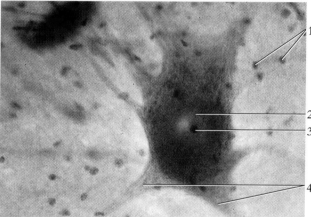

Figure 3.41 400X
Neuron smear.
  1. Nuclei of surrounding neuroglial cells
  2. Nucleus of neuron
  3. Nucleolus of neuron
  4. Dendrites of neuron

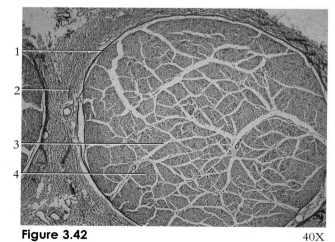

Figure 3.42 40X
Cross section of a nerve.
  1. Perineurium   3. Endoneurium
  2. Epineurium   4. Bundle of axons

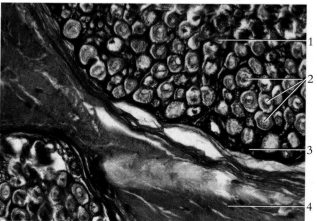
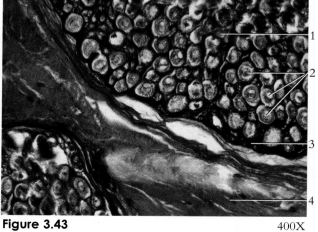

Figure 3.43 400X
Cross section of a nerve.
  1. Endoneurium   3. Perineurium
  2. Axons   4. Epineurium

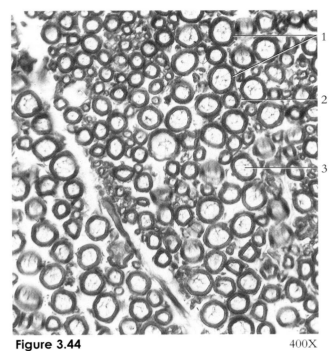

Figure 3.44   400X
A nerve stained with osmium.
1. Myelin sheath
2. Endoneurium
3. Axon

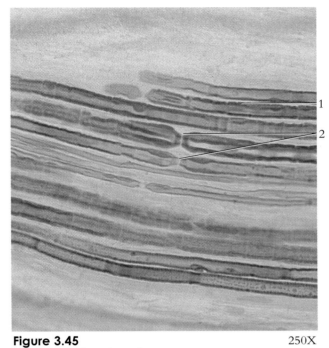

Figure 3.45   250X
A longitudinal section of axons.
1. Myelin sheath
2. Neurofibril nodes (nodes of Ranvier)

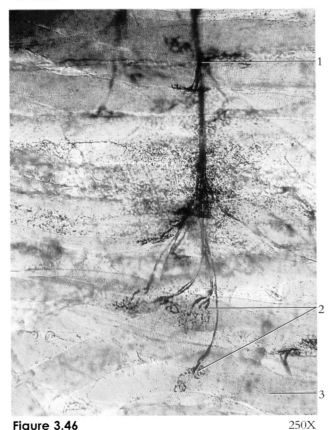

Figure 3.46   250X
Neuromuscular junction.
1. Motor nerve   3. Skeletal muscle
2. Motor end plates   fiber

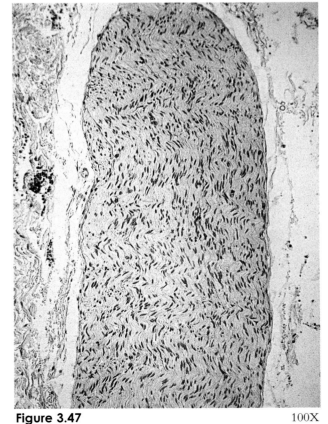

Figure 3.47   100X
Cross section of unmyelinated nerve.

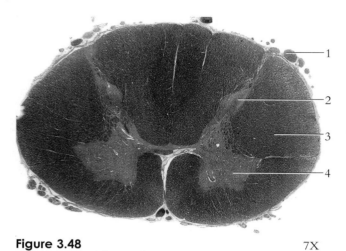

Figure 3.48   7X
Cross section of the spinal cord.
1. Posterior (dorsal) root of spinal nerve
2. Posterior (dorsal) horn (gray matter)
3. Spinal cord tract (white matter)
4. Anterior (ventral) horn (gray matter)

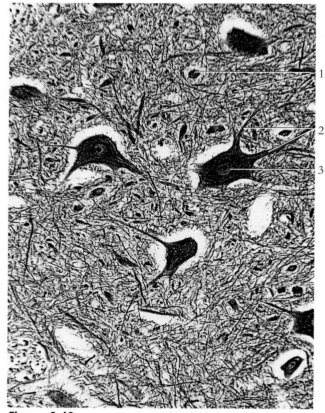

Figure 3.49   250X
Motor neurons from spinal cord.
1. Neuroglia cells
2. Dendrites
3. Nucleus

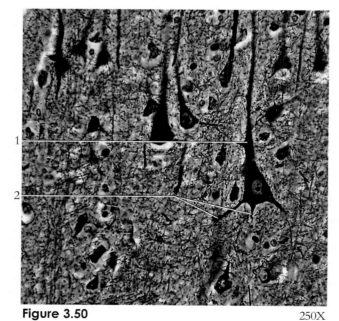

Figure 3.50   250X
Pyramidal neurons from the cerebral cortex.
1. Apical dendrite
2. Other dendrites

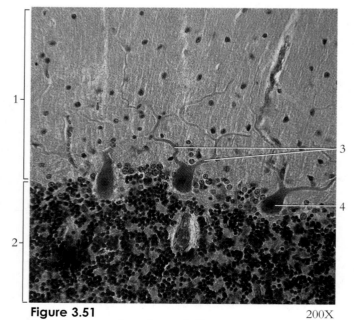

Figure 3.51   200X
Purkinje neurons from the cerebellum.
1. Molecular layer of cerebellar cortex
2. Granular layer of cerebellar cortex
3. Dendrites of Purkinje cell
4. Purkinje cell body

The integumentary system consists of the **integument**, or **skin**, and its associated **hair**, **glands**, and **nails** (fig. 4.1). The skin is composed of an outer **epidermis** consisting of four or five layers and a **dermis** consisting of two layers. The **hypodermis (subcutaneous tissue)** connects the skin to the underlying organs.

The stratified squamous epithelium of the epidermis is divisible into five **strata**, or layers. From superficial to deep, they are the **stratum corneum**, the **stratum lucidum** (only in skin of the palms and soles), the **stratum granulosum**, the **stratum spinosum**, and the **stratum basale**. The strata basale and spinosum undergo mitosis (cell division) and are collectively called the **stratum germinativum**. Pigments are found in the stratum germinativum and the protein **keratin** is found in all but the deepest epidermal layers. Both are protective. The stratum corneum is **cornified** (hardened and scale-like) for further protection.

The dermis is divisible into the **stratum papillarosum** (papillary layer) and the **stratum reticularosum** (reticular layer). The hypodermis is the deep, binding layer of connective tissue.

The skin provides several important functions, including: 1) protection of the body from disease and external injury. Keratin and an acidic oily secretion on the surface protect the skin from water and microorganisms. Cornification protects against abrasion, and **melanin** (a dark pigment) is a barrier to UV light; 2) regulation of body fluids and temperatures by radiation, convection, and the antagonistic effects of sweating and shivering; 3) permits the absorption of some UV light, respiratory gases, steroids, and fat-soluble vitamins; 4) synthesizes melanin and keratin, which remain in the skin, and vitamin D, which is used elsewhere in the body; 5) sensory reception provided through cutaneous receptors throughout the dermis and hypodermis; and 6) development and growth of hair and certain exocrine glands.

Formed prenatally as invaginations of the epidermis into the dermis, hair, glands and nails provide protection to the skin. Each hair develops in a **hair follicle** and is protective against sunlight and mild abrasions. Integumentary glands are classified as **sebaceous** (oil secreting), **sudoriferous** (sweat), and **ceruminous** (wax-producing). (**Mammary glands** are specialized sweat glands that produce milk in a lactating female.) A nail protects the terminal end of each digit. The fingernails also aid in picking up objects and scratching.

Figure 4.1

The skin and certain epidermal structures.

1. Epidermis
2. Dermis
3. Hypodermis
4. Shaft of hair
5. Stratum corneum
6. Stratum basale
7. Sweat duct
8. Sensory receptor
9. Sweat duct
10. Sebaceous gland
11. Arrector pili muscle
12. Hair follicle
13. Apocrine sweat gland
14. Eccrine sweat gland
15. Bulb of hair
16. Adipose tissue
17. Cutaneous blood vessels

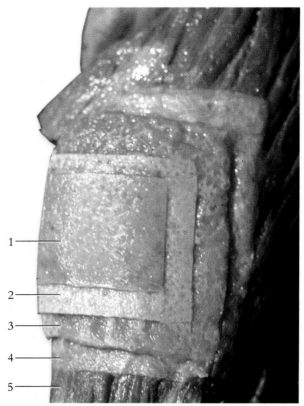

Figure 4.2
The gross structure of the skin and underlying fascia.
1. Epidermis   4. Fascia
2. Dermis   5. Muscle
3. Hypodermis

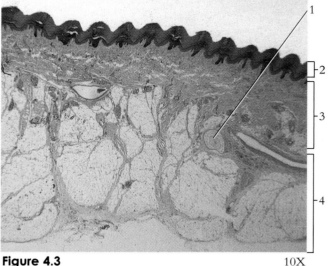

Figure 4.3   10X
Skin.
1. Lamellated (Pacinian) corpuscle   3. Dermis
2. Epidermis   4. Hypodermis

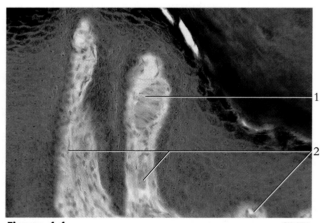

Figure 4.4   200X
Corpuscle of touch.
1. Corpuscle of touch (Meissner's corpuscle)
2. Dermal papillae

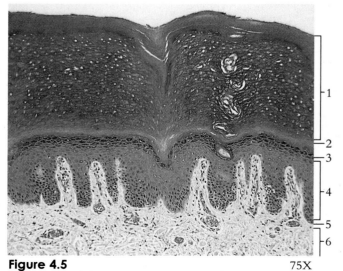

Figure 4.5   75X
Epidermis and dermis.
1. Stratum corneum   4. Stratum spinosum
2. Stratum lucidum   5. Stratum basale
3. Stratum granulosum   6. Dermis

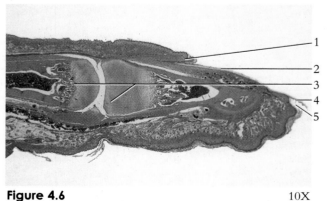

Figure 4.6   10X
Fingertip.
1. Eponychium   4. Free border of nail
2. Nail plate   5. Hyponychium
3. Phalanges

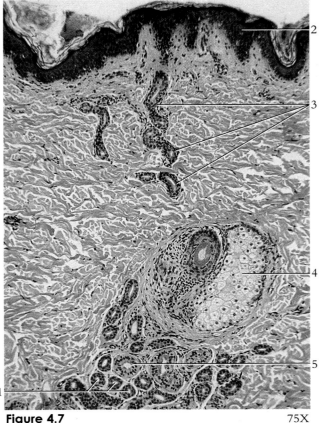

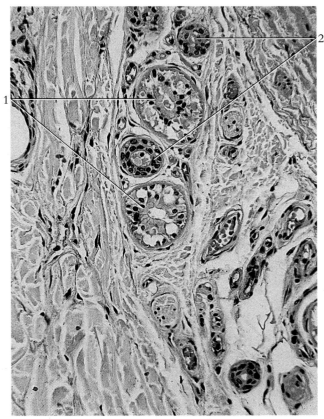

Figure 4.7   75X
Sweat gland.
1. Excretory portion of sweat gland
2. Epidermis
3. Excretory duct of sweat gland (coiling toward surface)
4. Sebaceous gland
5. Secretory portion of sweat gland

Figure 4.8   200X
Sweat gland.
1. Secretory portion
   (large diameter with light–staining columnar cells)
2. Excretory portion
   (small diameter with dark–staining stratified cuboidal cells)

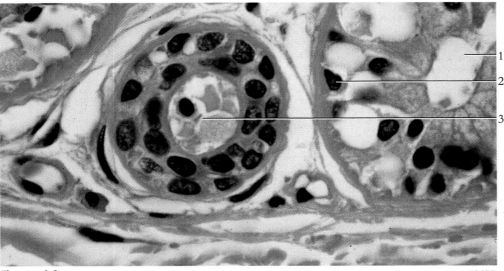

Figure 4.9   600X
Sweat gland.
1. Lumen of secretory portion
2. Myoepithelial cell
3. Lumen of excretory portion

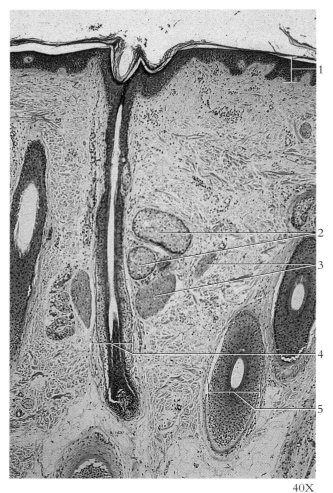

40X

Figure 4.10
Hair follicle.
1. Epidermis
2. Sebaceous glands
3. Arrector pili muscle
4. Hair follicle
5. Hair follicle (oblique cut)

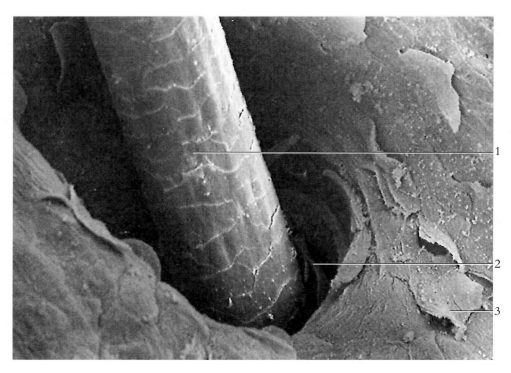

Figure 4.11
An electron micrograph of a hair emerging from a hair follicle.
1. Shaft of hair
   (note the scale–like pattern)
2. Hair follicle
3. Epithelial cell from
   stratum corneum

The skeletal system of an adult human is composed of approximately 206 bones—the number varies from person to person depending on genetic variations. Some adults have extra bones in the skull called **sutural (wormian) bones**. Additional bones may develop in tendons as the tendons move across a joint. Bones formed this way are called **sesamoid bones**, and the patella (kneecap) is an example.

The skeleton is divided into axial and appendicular portions (table 5.1). The **axial skeleton** consists of the bones that form the axis of the body and that support and protect the organs of the head, neck, and trunk. The axial skeleton includes the bones of the skull, auditory ossicles, hyoid bone, vertebral column, and rib cage.

The **appendicular skeleton** (see chapter 6) is composed of the bones of the upper and lower extremities and the bony gir-dles, which anchor the appendages to the axial skeleton. The appendicular skeleton includes the bones of the pectoral girdle, upper extremities, pelvic girdle, and lower extremities.

The mechanical functions of the bones of the skeleton include the support and protection of softer body tissues and organs. Also, certain bones function as levers during body movement. The metabolic functions of bones include **hemopoiesis**, or manufacture of blood cells, and mineral storage. Calcium and phosphorus are the two principal minerals stored within bone, and give bone its rigidity and strength.

The bones of the skeleton are classified into four principal types on the basis of shape rather than size. The four classes of bones are **long bones, short bones, flat bones** and **irregular bones** (fig. 5.1).

Table 5.1 Classifications of the bones of the adult skeleton.

Axial Skeleton	Rib cage	Skull - 22 bones	parietal bone (2)
	25 bones	14 facial bones	occipital bone (1)
	rib (24)	maxilla (2)	temporal bone (2)
	sternum (1)	palatine bone (2)	sphenoid bone (1)
	Vertebral column	zygomatic bone (2)	ethmoid bone (1)
	26 bones	lacrimal bone (2)	**Auditory ossicles—**
	cervical vertebra (7)	nasal bone (2)	6 bones
	thoracic vertebra (12)	vomer (1)	malleus (2)
	lumbar vertebra (5)	inferior nasal concha (2)	incus (2)
	sacrum (1) (5 fused bones)	mandible (1)	stapes (2)
	coccyx (1) (3–5 fused bones)	8 cranial bones	**Hyoid**
		frontal bone (1)	1 bone
Appendicular Skeleton	**Pectoral girdle**	**Upper extremities**	**Lower extremities**
	4 bones	60 bones	60 bones
	scapula (2)	metacarpal bone (10)	femur (2)
	clavicle (2)	humerus (2)	tarsal bone (14)
	Pelvic girdle	carpal bone (16)	tibia (2)
	2 bones	radius (2)	metatarsal bone (10)
	os coxae (2) (each contains 3	metacarpal bone (10)	fibula (2)
	fused bones: ilium, ischium,	ulna (2)	phalanx (28)
	and pubis)	phalanx (28)	patella (2)

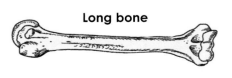

Long bone

Short bone

Flat bone

Irregular bone

Figure 5.1
Shapes of bones.

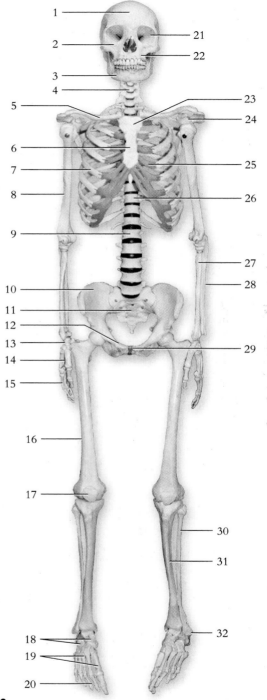

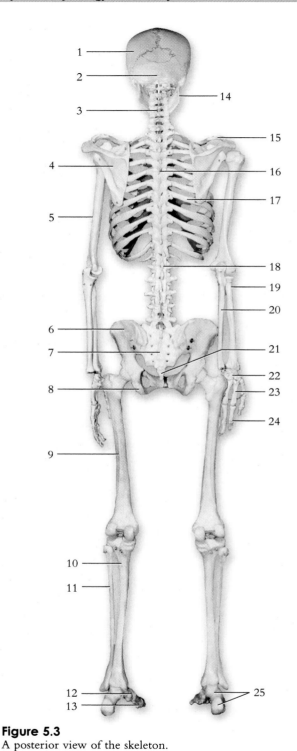

Figure 5.3
An anterior view of the skeleton.

1. Frontal bone
2. Zygomatic bone
3. Mandible
4. Cervical vertebra
5. Clavicle
6. Body of sternum
7. Rib
8. Humerus
9. Lumbar vertebra
10. Ilium
11. Sacrum
12. Pubis
13. Carpal bones
14. Metacarpal bones
15. Phalanges
16. Femur
17. Patella
18. Tarsal bones
19. Metatarsal bones
20. Phalanges
21. Orbit
22. Maxilla
23. Manubrium
24. Scapula
25. Costal cartilage
26. Thoracic vertebra
27. Radius
28. Ulna
29. Symphysis pubis
30. Fibula
31. Tibia
32. Calcaneus

Figure 5.3
A posterior view of the skeleton.

1. Parietal bone
2. Occipital bone
3. Cervical vertebra
4. Scapula
5. Humerus
6. Ilium
7. Sacrum
8. Ischium
9. Femur
10. Tibia
11. Fibula
12. Metatarsal bones
13. Phalanges
14. Mandible
15. Clavicle
16. Thoracic vertebra
17. Rib
18. Lumbar vertebra
19. Radius
20. Ulna
21. Coccyx
22. Carpal bones
23. Metacarpal bones
24. Phalanges
25. Tarsal bones

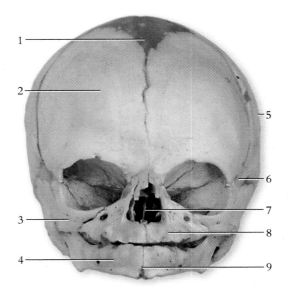

Figure 5.4
An anterior view of the fetal skull.
1. Anterior fontanel
2. Frontal bone
3. Zygomatic bone
4. Mandible
5. Parietal bone
6. Anterolateral fontanel
7. Nasal septum
8. Maxilla
9. Mental symphysis

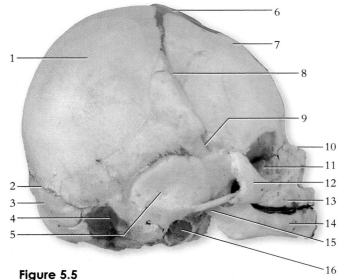

Figure 5.5
A lateral view of the fetal skull.
1. Parietal bone
2. Area of lambdoidal suture
3. Occipital bone
4. Posterolateral fontanel
5. Temporal bone
6. Anterior fontanel
7. Frontal bone
8. Area of coronal suture
9. Anterolateral fontanel
10. Nasal bone
11. Sphenoid bone
12. Zygomatic bone
13. Maxilla
14. Mandible
15. Condylar process
16. External acoustic meatus

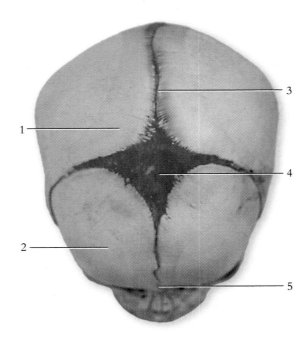

Figure 5.6
A superior view of the fetal skull.
1. Parietal bone
2. Frontal bone
3. Area of sagittal suture
4. Anterior fontanel
5. Synostosis

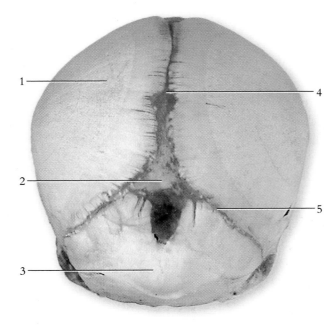

Figure 5.7
A posterior view of the fetal skull.
1. Parietal bone
2. Posterior fontanel
3. Occipital bone
4. Area of sagittal suture
5. Area of lambdoidal suture

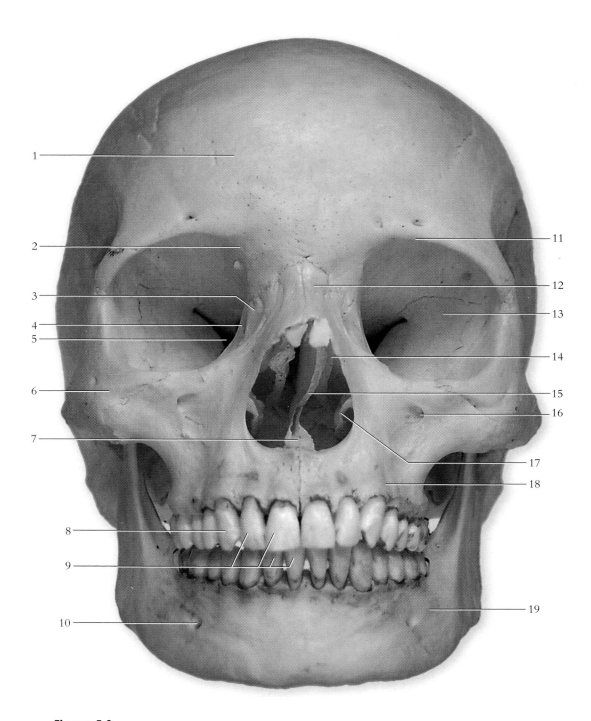

Figure 5.8

An anterior view of the skull.

1. Frontal bone
2. Supraorbital foramen
3. Lacrimal bone
4. Orbital plate of
   ethmoid bone
5. Superior
   orbital fissure
6. Zygomatic bone

7. Vomer
8. Canine
9. Incisors
10. Mental foramen
11. Supraorbital margin
12. Nasal bone
13. Sphenoid bone
14. Middle nasal concha of
    ethmoid bone

15. Perpendicular plate
    of ethmoid bone
16. Infraorbital foramen
17. Inferior nasal concha
18. Maxilla
19. Mandible

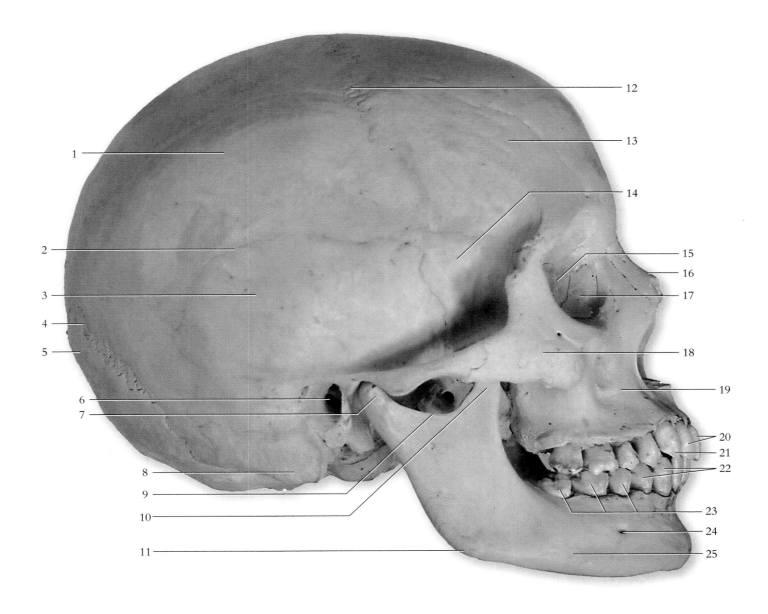

Figure 5.9

A lateral view of the skull.

1. Parietal bone
2. Squamosal suture
3. Temporal bone
4. Lambdoidal suture
5. Occipital bone
6. External acoustic meatus
7. Condylar process of mandible
8. Mastoid process of temporal bone

9. Mandibular notch
10. Coronoid process of mandible
11. Angle of mandible
12. Coronal suture
13. Frontal bone
14. Sphenoid bone
15. Orbital plate of ethmoid bone
16. Nasal bone

17. Lacrimal bone
18. Zygomatic bone
19. Maxilla
20. Incisors
21. Canine
22. Premolars
23. Molars
24. Mental Foramen
25. Mandible

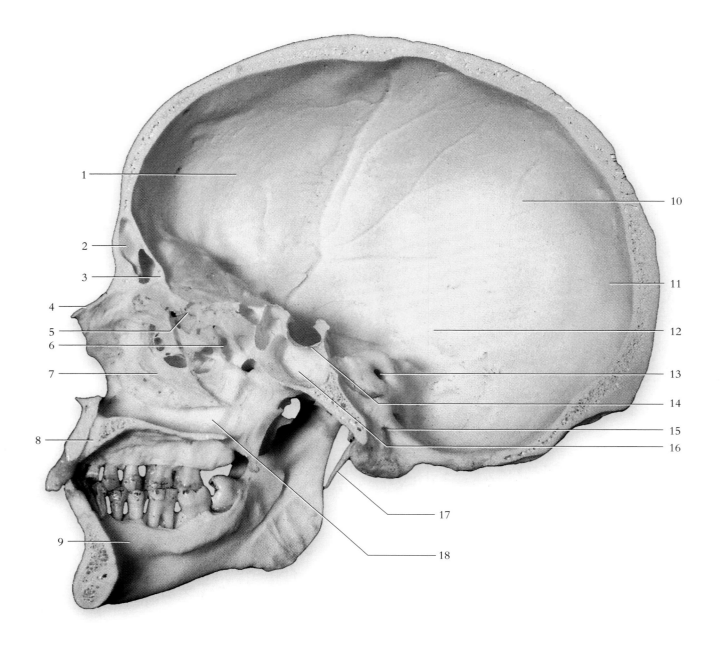

Figure 5.10

A sagittal view of the skull.

1. Frontal bone
2. Frontal sinus
3. Crista galli of
   ethmoid bone
4. Nasal bone
5. Cribriform plate
   of ethmoid bone

6. Ethmoidal sinus
7. Nasal concha
8. Maxilla
9. Mandible
10. Parietal bone
11. Occipital bone
12. Temporal bone

13. Internal acoustic meatus
14. Sella turcica
15. Hypoglossal canal
16. Sphenoidal sinus
17. Styloid process of
    temporal bone
18. Vomer

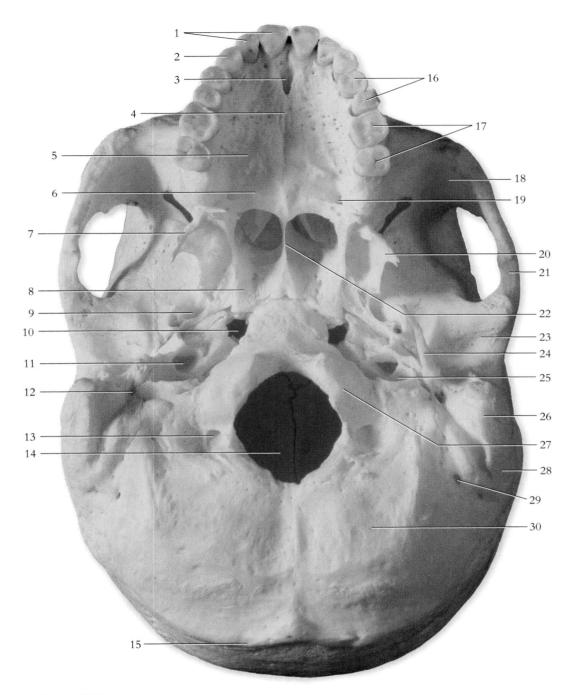

Figure 5.11
An inferior view of the skull.

1. Incisors
2. Canine
3. Incisive foramen
4. Median palatine suture
5. Maxilla
6. Palatine bone
7. Lateral pterygoid process of sphenoid bone
8. Sphenoid bone
9. Foramen ovale
10. Foramen lacerum
11. Carotid canal
12. Stylomastoid foramen
13. Condyloid canal
14. Foramen magnum
15. Superior nuchal line
16. Premolars
17. Molars
18. Zygomatic bone
19. Greater palatine foramen
20. Medial pterygoid process of sphenoid bone
21. Zygomatic arch
22. Vomer
23. Mandibular fossa
24. Styloid process of temporal bone
25. Jugular fossa
26. Mastoid process of temporal bone
27. Occipital condyle
28. Temporal bone
29. Mastoid foramen
30. Occipital bone

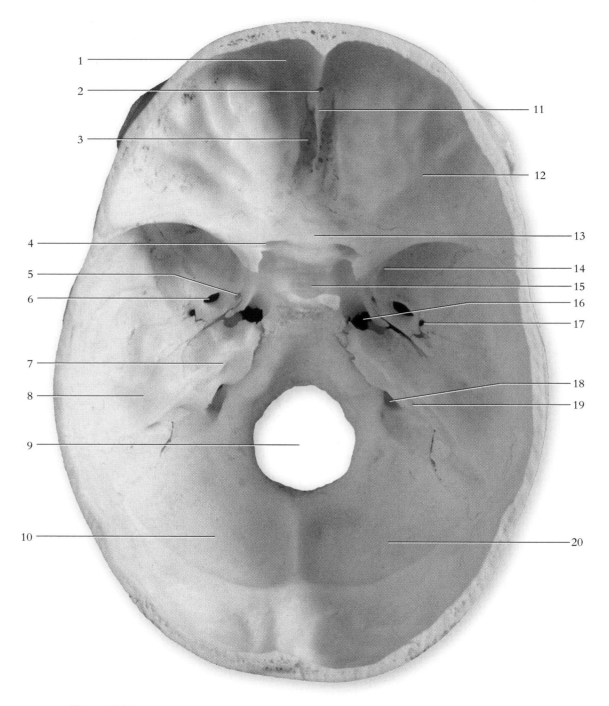

Figure 5.12

A superior view of the cranium.

1. Frontal bone
2. Foramen cecum
3. Cribriform plate of ethmoid bone
4. Optic canal
5. Foramen rotundum
6. Foramen ovale
7. Petrous part of temporal bone
8. Temporal bone
9. Foramen magnum
10. Occipital bone
11. Crista galli of ethmoid bone
12. Anterior cranial fossa
13. Sphenoid bone
14. Foramen rotundum
15. Sella turcica of sphenoid bone
16. Foramen lacerum
17. Foramen spinosum
18. Jugular foramen
19. Internal acoustic meatus
20. Posterior cranial fossa

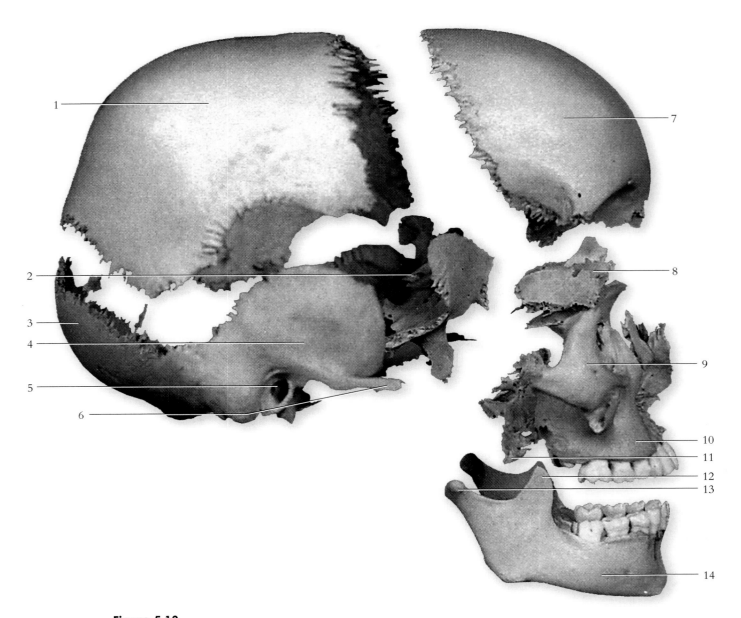

Figure 5.13
A lateral view of a disarticulated skull.

1. Parietal bone
2. Sphenoid bone
3. Occipital bone
4. Temporal bone
5. External acoustic
   meatus

6. Zygomatic process
   of temporal bone
7. Frontal bone
8. Ethmoid bone
9. Zygomatic bone
10. Maxilla

11. Palatine bone
12. Coronoid process
    of mandible
13. Condylar process
    of mandible
14. Mandible

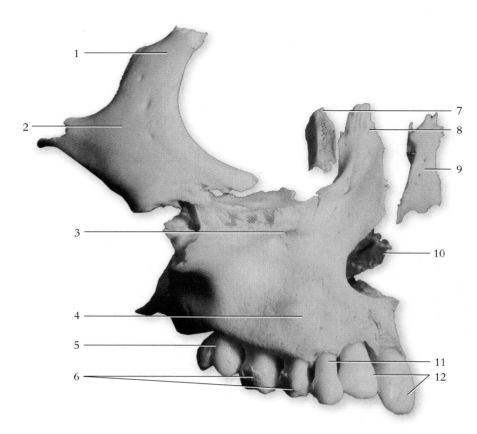

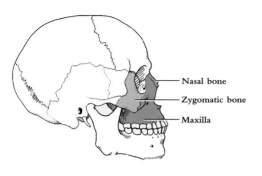

Figure 5.14
Bones of the left facial region.
1. Orbital process of zygomatic bone
2. Zygomatic bone
3. Infraorbital foramen
4. Maxilla
5. Molar
6. Premolars
7. Lacrimal bone
8. Frontal process of maxilla
9. Nasal bone
10. Inferior nasal concha
11. Canine
12. Incisors

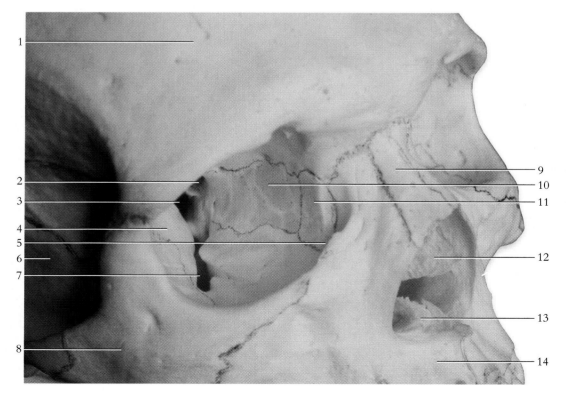

Figure 5.15
The orbit of the skull.
1. Frontal bone
2. Optic foramen
3. Superior orbital fissure
4. Sphenoid bone
5. Lacrimal foramen
6. Temporal bone
7. Inferior orbital fissure
8. Zygomatic bone
9. Nasal bone
10. Ethmoid bone
11. Lacrimal bone
12. Inferior nasal concha
13. Vomer
14. Maxilla

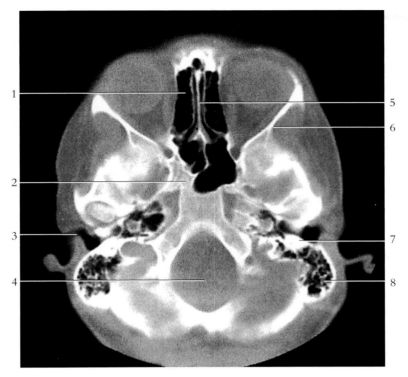

Figure 5.16
A CT transaxial section through the skull.
 1. Nasal cavity
 2. Sphenoid bone
 3. External acoustic meatus
 4. Foramen magnum
 5. Nasal septum
 6. Wall of bony orbit
 7. Middle-ear chamber
 8. Mastoidal sinus

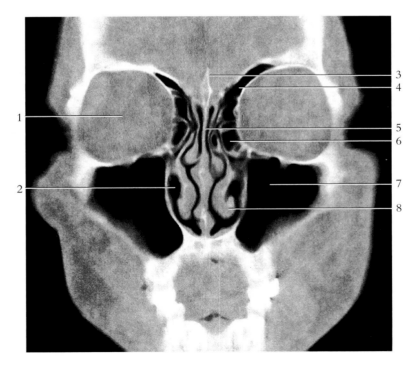

Figure 5.17
A CT image of the nasal cavity and paranasal sinuses.
 1. Eyeball in orbit
 2. Nasal cavity
 3. Crista galli
 4. Frontal sinus
 5. Perpendicular plate of ethmoid bone
 6. Ethmoidal sinus
 7. Maxillary sinus
 8. Inferior nasal concha

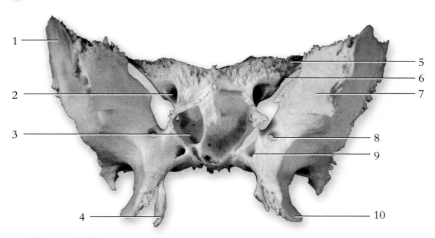

Figure 5.18

An anterior view of the sphenoid bone.

1. Greater wing of sphenoid bone
2. Optic foramen
3. Opening into sphenoidal sinus
4. Medial pterygoid plate
5. Lesser wing of sphenoid bone
6. Superior orbital fissure
7. Orbital surface of greater wing of sphenoid bone
8. Foramen rotundum
9. Pterygoid canal
10. Lateral pterygoid plate

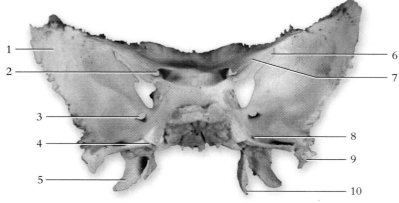

Figure 5.19

A posterior view of the sphenoid bone.

1. Greater wing of sphenoid bone
2. Optic foramen
3. Foramen rotundum
4. Pterygoid canal
5. Lateral pterygoid plate
6. Superior orbital fissure
7. Lesser wing of sphenoid bone
8. Foramen ovale
9. Spine of sphenoid bone
10. Medial pterygoid plate

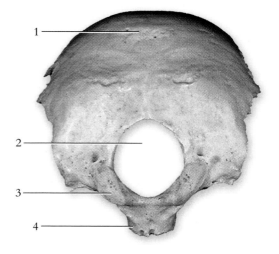

Figure 5.20

An inferior view of the occipital bone.

1. External occipital protuberance
2. Foramen magnum
3. Occipital condyle
4. Pharyngeal tubercle

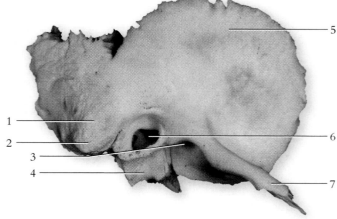

Figure 5.21

A lateral view of the temporal bone.

1. Mastoid part of temporal bone
2. Mastoid process
3. Mandibular fossa
4. Tympanic part of temporal bone
5. Squamous part of temporal bone
6. External acoustic meatus
7. Zygomatic process of temporal bone

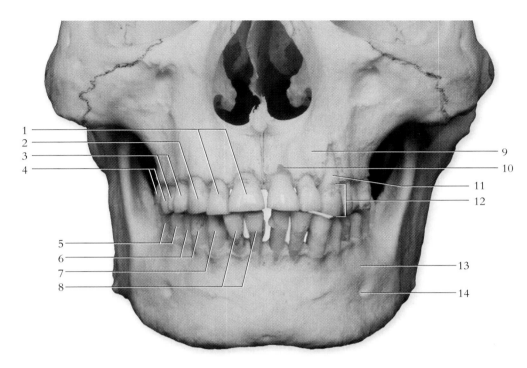

Figure 5.22
An anterior view of the jaws
and teeth.
  1. Superior incisors
  2. Superior canine
  3. Superior premolars
  4. Superior molars
  5. Inferior molars
  6. Inferior premolars
  7. Inferior canine
  8. Inferior incisors
  9. Maxilla
  10. Dental alveolus
  11. Root of tooth
  12. Crown of tooth
  13. Mandible
  14. Mental foramen

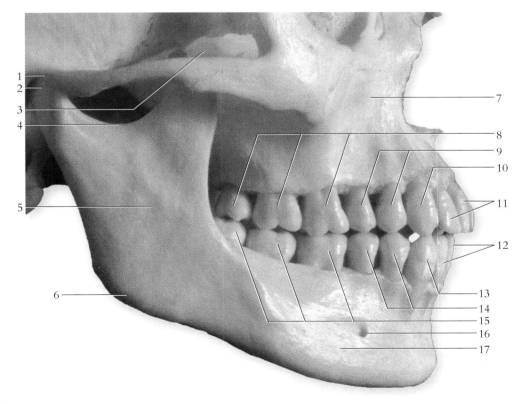

Figure 5.23
A lateral view of the jaws
and teeth.
  1. Temporomandibular joint
  2. Condylar process of
     mandible
  3. Coronoid process of
     mandible
  4. Mandibular notch
  5. Ramus of mandible
  6. Angle of mandible
  7. Maxilla
  8. Superior molars
  9. Superior premolars
  10. Superior canine
  11. Superior incisors
  12. Inferior incisors
  13. Inferior canine
  14. Inferior premolars
  15. Inferior molars
  16. Mental foramen
  17. Body of mandible

Figure 5.24
Eruption of teeth seen in a dissected skull of a youth (9 to 12 years old).
1. Permanent second molar
2. Permanent first molar
3. Permanent first molar
4. Permanent third molar
5. Permanent second molar
6. Deciduous second premolar
7. Permanent second premolar
8. Permanent second premolar
9. Permanent canine
10. Deciduous second premolar
11. Permanent first premolar
12. Deciduous canine
13. Incisors
14. Deciduous canine
15. Deciduous first premolar
16. Permanent canine
17. Permanent first premolar

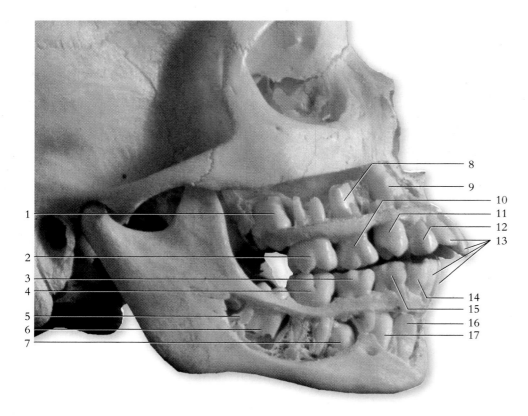

Figure 5.25
A medial view of the jaws and teeth.
1. Ethmoidal sinus
2. Inferior nasal concha
3. Inferior meatus
4. Hard palate
5. Maxilla
6. Superior molars
7. Superior premolars
8. Incisors
9. Inferior premolars
10. Inferior molars
    (note impacted wisdom tooth)
11. Mandible
12. Sphenoidal sinus
13. Styloid process of temporal bone
14. Medial pterygoid process

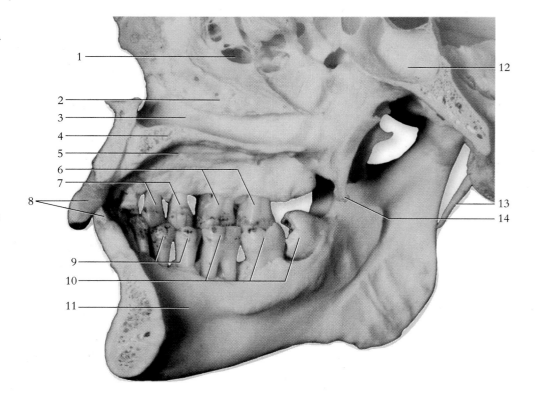

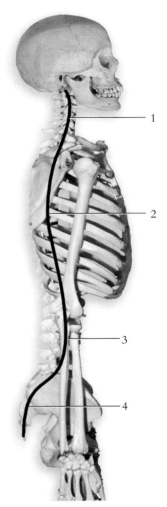

Figure 5.26
The curvature of the
vertebral column.
1. Cervical curvature
2. Thoracic curvature
3. Lumbar curvature
4. Pelvic (sacral) curvature

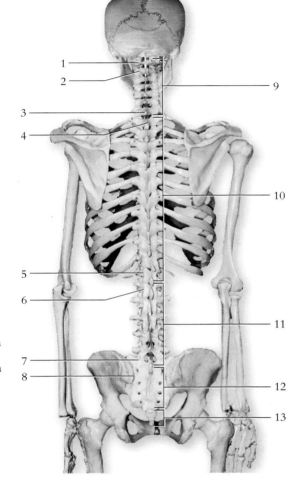

Figure 5.27
A posterior view of the
vertebral column.
1. Atlas
2. Axis
3. Seventh cervical vertebra
4. First thoracic vertebra
5. Twelfth thoracic vertebra
6. First lumbar vertebra
7. Fifth lumbar vertebra
8. Sacroiliac joint
9. Cervical vertebrae
10. Thoracic vertebrae
11. Lumbar vertebrae
12. Sacrum
13. Coccyx

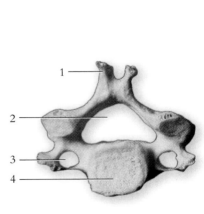

Cervical vertebra

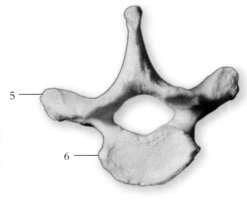

Thoracic vertebra

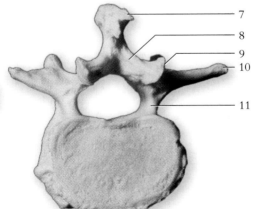

Lumbar vertebra

Figure 5.28
Representative vertebrae.

1. Spinous process (note that it is bifid)
2. Vertebral canal (in transverse process)
3. Transverse foramen
4. Body of vertebrae

5. Facet for head of rib
6. Inferior facet for head of rib
7. Spinous process
8. Lamina

9. Superior articular surface
10. Transverse process
11. Pedicle

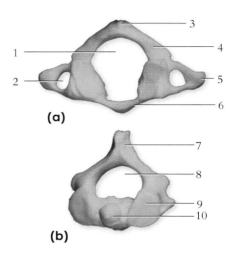

Figure 5.29

Superior views of (a) the atlas and (b) the axis.

1. Vertebral canal
2. Transverse foramen
3. Spinous process of atlas
4. Lamina of neural arch
5. Transverse process of atlas
6. Anterior arch of atlas
7. Spinous process of axis
8. Vertebral canal
9. Superior articular facet for atlas
10. Dens (odontoid process)

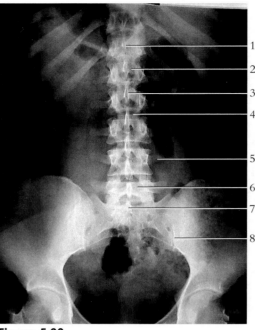

Figure 5.30

A radiograph of the lumbar vertebrae.

1. T12
2. Body of L1
3. Spinous process of L2
4. Intervertebral disc
5. Transverse process of L4
6. Lamina of L5
7. Sacrum
8. Sacroiliac joint

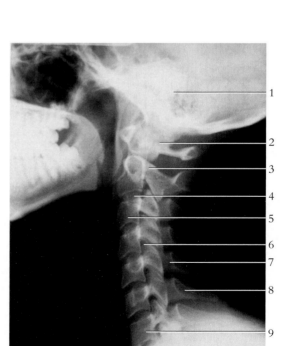

Figure 5.31

A radiograph of the cervical vertebrae.

1. Occipital condyle
2. Atlas
3. Axis
4. Intervetebral disc
5. Body of C3
6. Intervetebral foramen
7. Spinous process of C5
8. Spinous process of C6
9. Body of C7

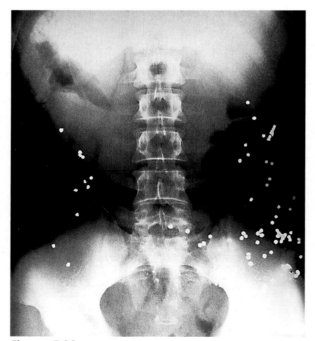

Figure 5.32

A radiograph of the lumbar region showing the locations of pellets from a shotgun wound.

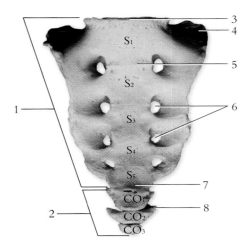

Figure 5.33
An anterior view of the sacrum and coccyx.

1. Sacrum
2. Coccyx
3. Base of sacrum
4. Ala
5. Transverse line
6. Anterior sacral foramina
7. Apex of sacrum
8. Coccygeal cornu

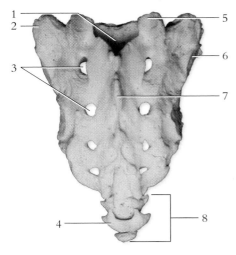

Figure 5.34
A posterior view of the sacrum and coccyx.

1. Superior portion of sacral canal
2. Ala
3. Posterior sacral foramina
4. Coccygeal cornu
5. Superior articular process
6. Auricular surface
7. Median sacral crest
8. Coccyx

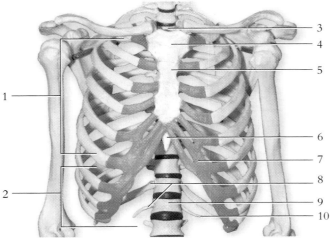

Figure 5.35
An anterior view of the rib cage.

1. True ribs (seven pairs)
2. False ribs (five pairs)
3. Jugular notch
4. Manubrium
5. Body of sternum
6. Xiphoid process
7. Costal cartilage
8. Floating ribs (inferior two pairs of false ribs)
9. Twelfth thoracic vertebra
10. Twelfth rib

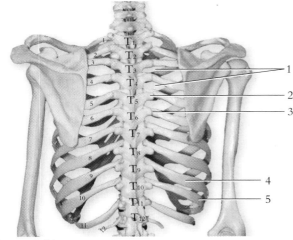

Figure 5.36
A posterior view of the rib cage.

1. Intercostal spaces
2. Transverse process of thoracic vertebra
3. Head of rib
4. Angle of rib
5. Body of rib

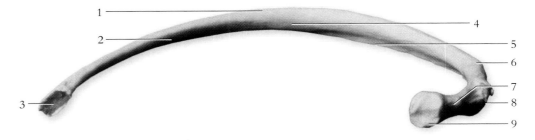

Figure 5.37
A .

1. Superior border
2. Body of rib
3. Articulation site for costal cartilage
4. Internal surface
5. Costal groove
6. Angle of rib
7. Neck of rib
8. Articular surface
9. Head of rib

The structure of the **pectoral girdle** and **upper extremities** of the appendicular skeleton is adaptive for freedom of movement and extensive muscle attachment. The structure of the **pelvic girdle** and **lower extremities** is adaptive for support and locomotion.

The pectoral girdle is composed of two **scapulae** and two **clavicles**. The clavicles attach the pectoral girdle to the axial skeleton at the **sternum**. The bones of each upper extremity are the **humerus**, **ulna**, **radius**, eight **carpal bones**, five **metacarpal bones**, and fourteen **phalanges**. In addition, two or more **sesamoid bones** at the interphalangeal joints are usually present.

The pelvic girdle, or **pelvis**, is formed by two **ossa coxae** (hip bones) which are united anteriorly by the **symphysis pubis**. The pelvic girdle is attached posteriorly to the sacrum of the vertebral column. Each **os coxae** consists of three separate bones: the **ilium**, the **ischium**, and the **pubis**. These bones are fused in an adult. The bones of each lower extremity are the **femur**, **patella**, **tibia**, **fibula**, seven **tarsal bones**, five **metatarsal bones**, and fourteen **phalanges**. In addition, two or more sesamoid bones at the interphalangeal joints are usually present.

The various surface features used to identify specific bones are presented in table 6.1.

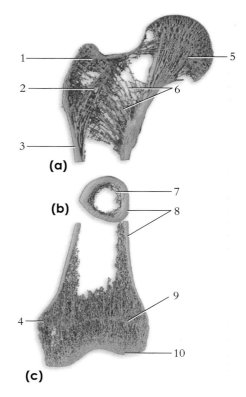

(a)

(b)

(c)

Figure 6.1
Sections of a human femur showing the outer compact bone and inner spongy bone. Note the trajectorial (stress) lines of the trabeculae of spongy bone.
(a) A coronal section of the proximal end of the femur
(b) A cross section of the body of the femur
(c) A coronal section of the distal end of the femur.
1. Greater trochanter
2. Spongy bone
3. Compact bone
4. Femoral epicondyle
5. Head of femur
6. Trabeculae of spongy bone
7. Spongy bone surrounding medullary cavity
8. Compact bone
9. Epiphyseal line (reminant of epiphyseal plate)
10. Condyle of femur

Table 6.1 Surface features used to identify specific bones.

Articulating surfaces	Description
Condyle	A large, rounded, articulating knob
Facet	A flattened or shallow articulating surface
Head	A prominent, rounded, articulating end of a bone
Nonarticulating prominences	
Crest	A narrow, ridgelike projection
Epicondyle	A projection above a condyle
Process	Any marked bony prominence
Spine	A sharp, slender process
Trochanter	A large process found only on the femur
Tubercle	A small rounded process
Tuberosity	A large roughened process
Depressions and openings	
Alveolus	A deep pit or socket
Fissure	A narrow, slitlike opening
Foramen (plural, foramina)	A rounded opening through a bone
Fossa	A flattened or shallow surface
Fovea	A small pit or depression; some are articular and some are not
Meatus, or canal	A tubelike passageway through a bone
Sinus	A cavity or hollow space in a bone
Sulcus	A groove in a bone

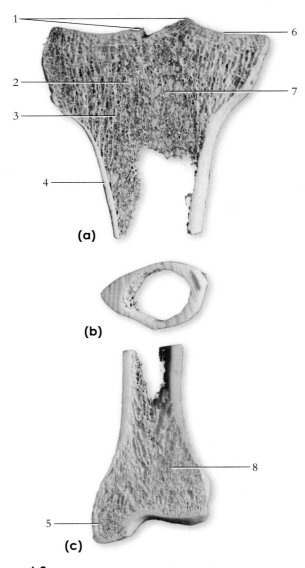

(a)

(b)

(c)

Figure 6.2
Sections of a human tibia showing the outer compact bone and the inner spongy bone. Note the trajectorial stress lines of the trabeculae of spongy bone.
(a) A coronal section of the proximal end of the tibia
(b) A cross section of the body of tibia
(c) A coronal section of the distal end of the tibia.

1. Intercondylar eminences
2. Epiphyseal line
3. Spongy bone
4. Compact bone
5. Medial malleolus
6. Lateral condyle
7. Trabeculae of spongy bone
8. Spongy bone

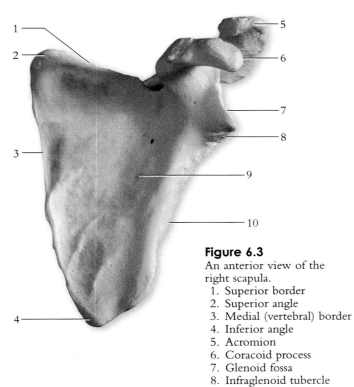

Figure 6.3
An anterior view of the right scapula.
1. Superior border
2. Superior angle
3. Medial (vertebral) border
4. Inferior angle
5. Acromion
6. Coracoid process
7. Glenoid fossa
8. Infraglenoid tubercle
9. Subscapular fossa
10. Lateral (axillary) border

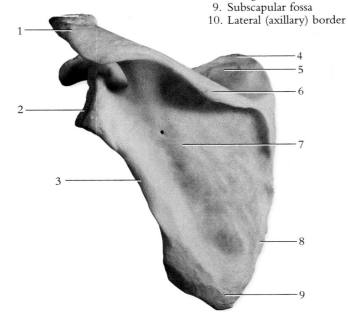

Figure 6.4
A posterior view of the right scapula.
1. Acromion
2. Glenoid fossa
3. Lateral (axillary) border
4. Superior angle
5. Supraspinous fossa
6. Spine
7. Infraspinous fossa
8. Medial (vertebral) border
9. Inferior angle

Figure 6.5
An inferior view of the clavicle.
1. Conoid tubercle
2. Acromial extremity
3. Sternal extremity

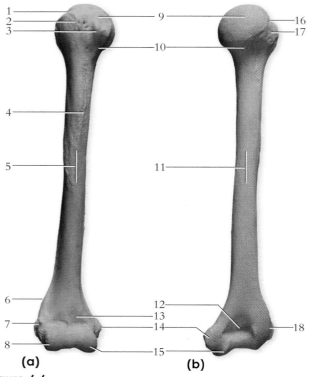

(a) **(b)**

Figure 6.6

The right humerus. (a) Anterior view (b) Posterior view.
1. Greater tubercle
2. Intertubercular groove
3. Lesser tubercle
4. Deltoid tuberosity
5. Anterior body (shaft) of humerus
6. Lateral supracondylar ridge
7. Lateral epicondyle
8. Capitulum
9. Head of humerus
10. Surgical neck
11. Posterior body (shaft) of humerus
12. Olecranon fossa
13. Coronoid fossa
14. Medial epicondyle
15. Trochlea
16. Anatomical neck
17. Greater tubercle
18. Lateral epicondyle

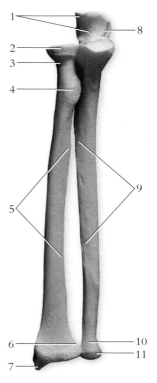

Figure 6.7

An anterior view of the right ulna and radius.
1. Trochlear notch
2. Head of radius
3. Neck of radius
4. Radial tuberosity
5. Interosseous margin
6. Ulnar notch of radius
7. Styloid process of radius
8. Olecranon
9. Interosseous margin
10. Neck of ulna
11. Head of ulna

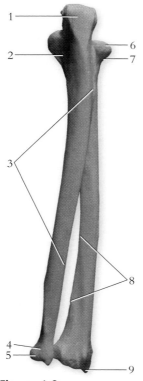

Figure 6.8

A posterior view of the right ulna and radius.
1. Olecranon
2. Radial notch of ulna
3. Interosseous margin
4. Head of ulna
5. Styloid process of ulna
6. Head of radius
7. Neck of radius
8. Interosseous margin
9. Styloid process of radius

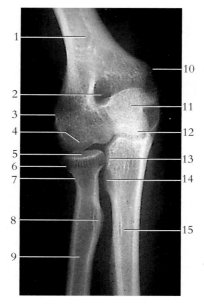

Figure 6.9

A radiograph of the left elbow region, posterior view.
1. Humerus
2. Olecranon fossa of humerus
3. Lateral epicondyle of humerus
4. Capitulum
5. Articular surface of radius
6. Head of radius
7. Neck of radius
8. Tuberosity of radius
9. Radius
10. Medial epicondyle of humerus
11. Olecranon
12. Trochlea
13. Radial notch of ulna
14. Tuberosity of ulna
15. Ulna

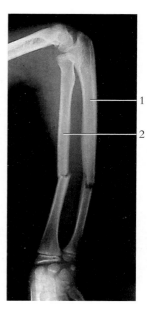

Figure 6.10 A radiograph showing fractures of the ulna and radius of a ten-year-old child. Notice the distal epiphyseal plates of the ulna and the radius.
1. Ulna
2. Radius

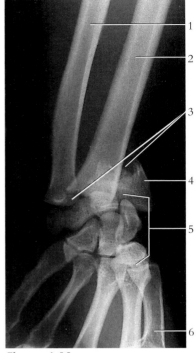

Figure 6.11
A radiograph showing a fracture of the distal portion of the radius and medial displacement of the antebrachium.
1. Ulna
2. Radius
3. Site of fracture
4. Styloid process of radius
5. Carpal bones
6. First metacarpal bone

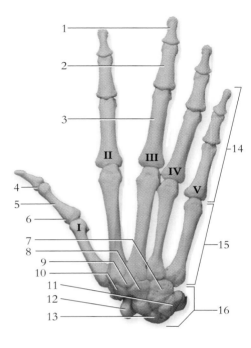

Figure 6.12
A posterior view of the right wrist and hand.
1. Distal phalanx of third digit
2. Middle phalanx of third digit
3. Proximal phalanx of third digit
4. Head of phalanx
5. Body of phalanx
6. Base of phalanx
7. Hamate bone
8. Capitate bone
9. Trapezoid bone
10. Trapezium bone
11. Triquetrum bone
12. Scaphoid bone
13. Lunate bone
14. Phalanges of fifth digit
15. Metacarpal bones (first-fifth)
16. Carpal bones

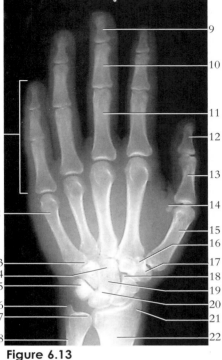

Figure 6.13
A radiograph of the right wrist and hand, anteroposterior view.
1. Phalanges of fifth digit
2. Fifth metacarpal bone
3. Hamate bone
4. Capitate bone
5. Triquetrum and pisiform bones
6. Styloid process of ulna
7. Distal radioulnar joint
8. Ulna
9. Distal phalanx of third digit
10. Middle phalanx of third digit
11. Proximal phalanx of third digit
12. Distal phalanx of pollex (thumb)
13. Proximal phalanx of pollex (thumb)
14. Sesamoid bone
15. First metacarpal bone
16. Trapezoid bone
17. Trapezium bone
18. Scaphoid bone
19. Capitate bone
20. Lunate bone
21. Styloid process of radius
22. Radius

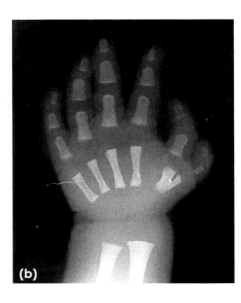

(a) (b)

Figure 6.14
A photograph (a) and a radiograph (b) showing polydactyly, having extra digits. Polydactyly is a common congenital deformity of the hand, although it also occurs in the foot. Notice that the carpal bones are still cartilaginous in a newborn and do not show up on a radiograph.

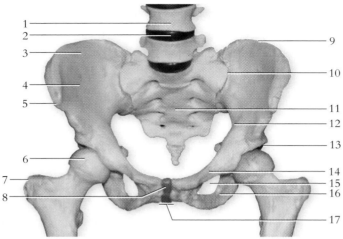

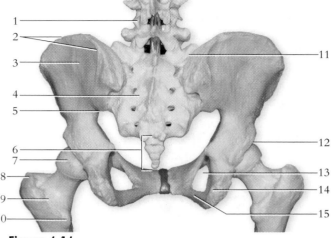

Figure 6.15

An anterior view of the articulated pelvic girdle showing the two coxal bones, the sacrum, and the two femora.

1. Lumbar vertebra
2. Intervertebral disc
3. Ilium
4. Iliac fossa
5. Anterior superior iliac spine
6. Head of femur
7. Greater trochanter
8. Symphysis pubis
9. Crest of the ilium
10. Sacroiliac joint
11. Sacrum
12. Pelvic brim
13. Acetabulum
14. Pubic crest
15. Obturator foramen
16. Ischium
17. Pubic angle

Figure 6.16

A posterior view of the articulated pelvic girdle showing the two coxal bones, the sacrum, and the two femora.

1. Lumbar vertebra
2. Crest of ilium
3. Ilium
4. Sacrum
5. Greater sciatic notch
6. Coccyx
7. Head of femur
8. Greater trochanter
9. Intertrochanteric crest
10. Lesser trochanter
11. Sacroiliac joint
12. Acetabulum
13. Obturator foramen
14. Ischium
15. Pubis

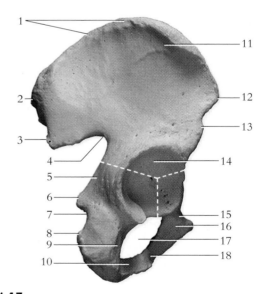

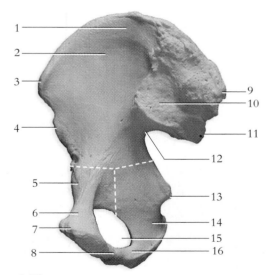

Figure 6.17

A lateral view of the right os coxae.

1. Crest of the ilium
2. Posterior superior iliac spine
3. Posterior inferior iliac spine
4. Greater sciatic notch
5. Ischial body
6. Ischial spine
7. Lesser sciatic notch
8. Ischial tuberosity
9. Ischium
10. Ischial ramus
11. Ilium
12. Anterior superior iliac spine
13. Anterior inferior iliac spine
14. Acetabulum
15. Superior ramus of pubis
16. Pubis
17. Obturator foramen
18. Inferior ramus of pubis

Figure 6.18

A medial view of the right os coxae.

1. Ilium
2. Iliac fossa
3. Anterior superior iliac spine
4. Anterior inferior iliac spine
5. Superior ramus of pubis
6. Pubis
7. Pubic crest
8. Pubic ramus
9. Posterior superior iliac spine
10. Auricular surface
11. Posterior inferior iliac spine
12. Greater sciatic notch
13. Ischial spine
14. Ischium
15. Obturator foramen
16. Ischial ramus

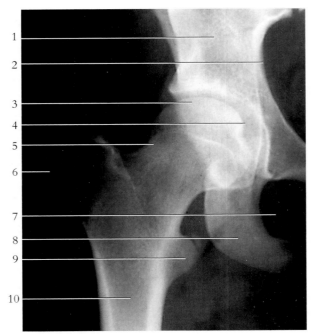

Figure 6.19
A radiograph of the left os coxae and
femur, anterior view.

1. Ilium
2. Pelvic brim
3. Head of femur
4. Acetabulum
5. Neck of femur
6. Greater trochanter
7. Obturator foramen
8. Ischium
9. Lesser trochanter
10. Femur

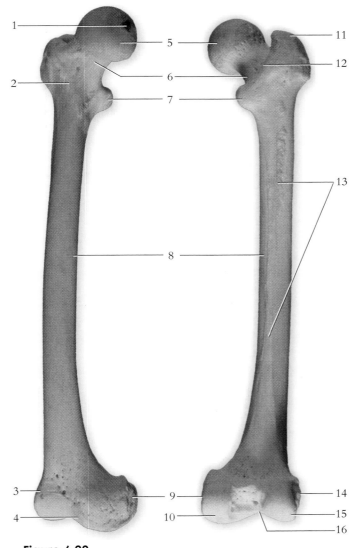

Figure 6.20
The right femur.

1. Fovea capitis femoris
2. Intertrochanteric line
3. Lateral epicondyle
4. Patellar surface
5. Head of femur
6. Neck of femur
7. Lesser trochanter
8. Body (shaft) of femur
9. Medial epicondyle
10. Medial condyle
11. Greater trochanter
12. Intertrochanteric crest
13. Linea aspera on body (shaft) of femur
14. Lateral epicondyle
15. Lateral condyle
16. Intercondylar fossa

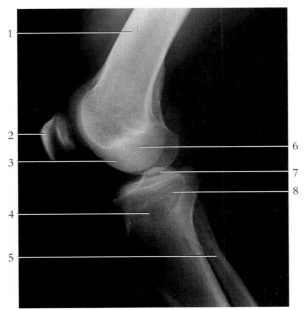

Figure 6.21
Radiograph of the right knee region, lateral view.

1. Femur
2. Patella
3. Condyle of femur
4. Tibia
5. Fibula
6. Lateral epicondyle of femur
7. Intercondylar eminences of tibia
8. Tibial tuberosity

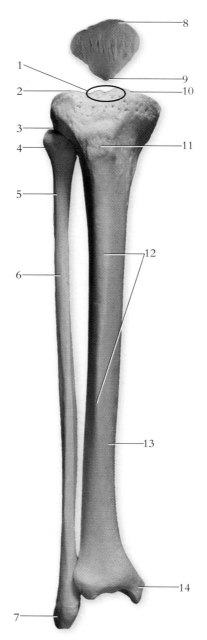

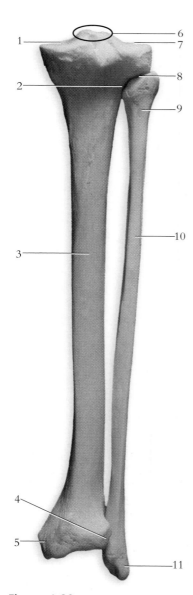

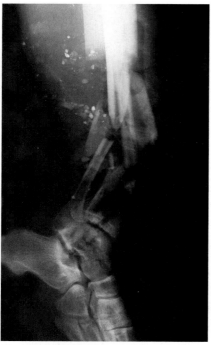

Figure 6.24
A severe fracture of the leg and ankle (talus bone). In this patient, the trauma was so extensive amputation of the leg was necessary.

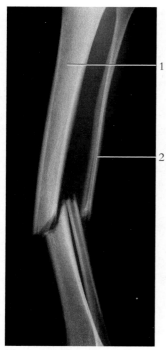

Figure 6.25
Radiograph showing fractures of the tibia and fibula.
1. Tibia
2. Fibula

Figure 6.22
An anterior view of the right patella, tibia, and fibula.
1. Intercondylar tubercles
2. Lateral condyle
3. Tibial articular facet of fibula
4. Head of fibula
5. Neck of fibula
6. Body (shaft) of fibula
7. Lateral malleolus
8. Base of patella
9. Apex of patella
10. Medial condyle
11. Tibial tuberosity
12. Anterior crest of tibia
13. Body (shaft) of tibia
14. Medial malleolus

Figure 6.23
A posterior view of the right tibia and fibula.
1. Medial condyle of tibia
2. Fibular articular facet of tibia
3. Body (shaft) of tibia
4. Fibular notch of tibia
5. Medial malleolus
6. Intercondylar tubercles
7. Lateral condyle of tibia
8. Head of fibula
9. Neck of fibula
10. Body (shaft) of fibula
11. Lateral malleolus

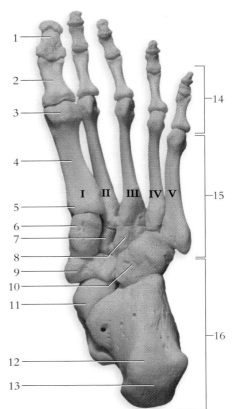

Figure 6.26
An inferior view of the left foot.
1. Distal phalanx of first digit
2. Proximal phalanx of first digit
3. Head of first metatarsal bone
4. Body of first metatarsal bone
5. Base of first metatarsal bone
6. Medial cuneiform
7. Intermediate cuneiform
8. Lateral cuneiform
9. Navicular bone
10. Cuboid bone
11. Talus bone
12. Calcaneus bone
13. Tuberosity of calcaneus
14. Phalanges of fifth digit
15. Metatarsal bones (first–fifth)
16. Tarsal bones

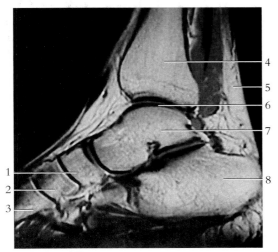

Figure 6.27
An MR image of the ankle region, medial view.
1. Navicular bone
2. Medial cuneiform
3. First metatarsal bone
4. Tibia
5. Tendo calcaneus
6. Tibiotalar joint
7. Talus
8. Calcaneus

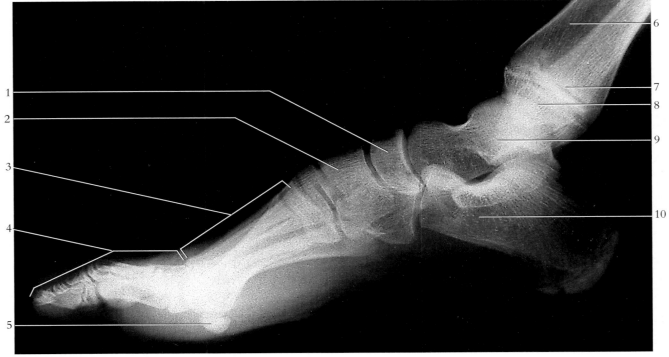

Figure 6.28
A radiograph of the right foot, medial view.
1. Navicular bone
2. First cuneiform
3. Metatarsal bones
4. Phalanges
5. Sesamoid bone
6. Tibia
7. Fibula (superimposed)
8. Tibiotalar joint
9. Talus
10. Calcaneus

⑦ Articulations

The structure of a joint determines its range of movement. Not all joints are flexible, however, and as one part of the body moves, other joints remain rigid to stabilize the body and maintain balance. **Arthrology** is the study of joints and **kinesiology** is the study of body movement, or the functional relationship between the skeleton, joints, muscles, and innervation (nerve supply) as they work together to produce coordinated movement.

The joints of the body are structurally classified into three basic kinds:

1. **Fibrous joints**
   Fibrous connective tissues join the skeletal structures; the joints lack a joint capsule, but in some slight movement is possible.
2. **Cartilaginous joints**
   Fibrocartilage or hyaline cartilage joins the skeletal structures; the joints lack a joint capsule, but in some, slight movement is possible.
3. **Synovial joints**
   Joint capsules containing synovial fluid are present between the articulating bones; articular cartilages and ligaments supporting the articulating bones are also present, which permit freedom of movement. Synovial joints are the freely moveable joints of the body.

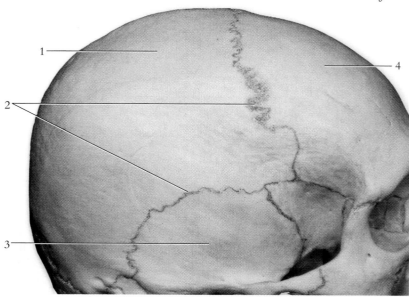

Figure 7.1
A Suture located between cranial bones is a type of fibrous joint.
1. Parietal bone
2. Sutures
3. Temporal bone
4. Frontal bone

Table 7.1 Movements permitted at synovial joints.

Type of movement	Description
Angular movement	Increase or decrease the joint angle
Flexion	Decreasing the angle between two bones
Extension	Increasing the angle between two bones
Hyperextension	Excessive extension beyond 180° (angle of anatomical position)
Dorsiflexion	Bending the foot toward the tibia
Plantar flexion	Bending the foot away from the tibia
Abduction	Movement of a body part away from the axis of the body, or away from the midsagittal plane, in a lateral direction
Inversion	Movement of the sole of the foot inward, or medially
Eversion	Movement of the sole of the foot outward, or laterally
Circular movement	
Rotation	Turning of a bone at joint axis
Circumduction	Movement of a body segment in a circular, conelike motion

Table 7.2 Types of joints.

Type	Structure and movement	Example
Fibrous	Fibrous connective tissue joins the skeletal structures	
Suture	Frequently serrated edges of articulating bones, seperated by a thin layer of fibrous tissue; no movement	Structures between bones, cranial bones
Syndesmoses	Articulating bones bound by an interosseous ligament; slight movement	Joints between tibia-fibula and radius-ulna
Gomphoses	Periodontal ligament binding teeth into dental aveoli of bone; no movement	Teeth secured in the dental alveoli (teeth sockets)
Cartilaginous	Fibrocartilage or hyaline cartilage joins the skeletal structures	
Symphyses	Thin pad of fibrocartilage between articulating bones; slight movement	Symphysis pubis, sacroiliac joint and intervertebral joints
Synchondroses	Mitotically active hyaline cartilage between skeletal structures; no movement	Epiphyseal plates between diaphysis and epiphyses of long bones
Synovial	Joint capsule between articulating bones, containing synovial fluid; extensive movement	
Gliding	Flattened or slightly curved articulating surfaces; sliding movement	Intercarpal and intertarsal joints
Hinge	Concave surface of one bone articulates with a depression of another; bending motion in one plane	Humeroulnar (elbow) and tibio-femoral (knee) joints; inter-phalangeal (finger) joints of digits
Pivot	Conical surface of one bone articulates with a depression of another; rotation about a central axis; rotational movement	Atlantoaxial joint; proximal radioulnar joint
Condyloid	Oval condyle of one bone articulates with elliptical cavity of another; biaxial movement	Radiocarpal (wrist) joint
Saddle	Concave and convex surface on each articulating bone; wide range of movement; biaxial movement	Carpometacarpal joint at the base of the thumb
Ball-and-socket	Rounded convex surface of one bone articulates with cuplike socket of another; movement in all planes and roation	Glenohumeral (shoulder) and coxal (hip) joints

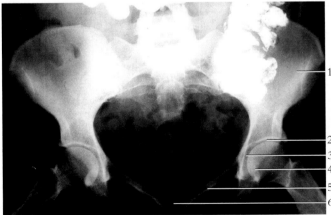

Figure 7.2
A symphsis is a type of cartilaginous joint. The symphysis pubis can be readily seen in this radiograph of the pelvic girdle. Note the head of the femur articulating with the acetabulum of the os coxae forming a ball-and-socket joint is freely moveable synovial joint.

1. Ilium
2. Acetabulum
3. Ball-and-socket joint
4. Head of femur
5. Pubis
6. Symphsis pubis

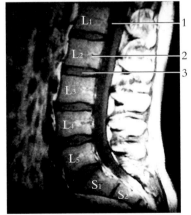

Figure 7.3
An intervertebral joint between vertebral bodies is a type of cartilaginous joint. The intervertebral discs of intervertebral joints can be readily seen in a lateral MR image of the lower back.

1. Spinal cord
2. Body of second lumbar vertebra (L2)
3. Intervertebral disc

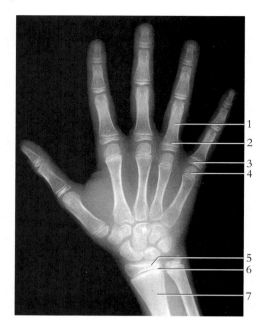

Figure 7.4
A synchondrosis is a type of cartilaginous joint located within a long bone at both the proximal and the distal epiphyseal plates as seen in a radiograph of a child's hand. Mitotic activity at a synchondrotic joint is responsible for linear (length) bone growth.
1. Diaphysis of phalanyx
2. Epiphyseal plate
3. Proximal epiphysis of phalanx
4. Distal epiphysis of metacarpal bone
5. Distal epiphysis of radius
6. Epiphyseal plate
7. Diaphysis of radius

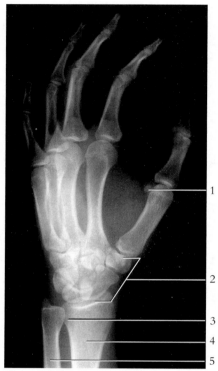

Figure 7.5
The side-to-side articulationof the ulna and radius forms a syndesmosis that is tightly bound by an interosseous ligament (not seen in radiograph). A syndesmosis is a type of fibrous joint that permits slight movement.

1. Sesamoid bone   4. Radius
2. Carpal bones   5. Ulna
3. Syndesmosis

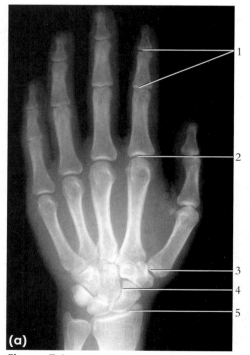

(a)

(b)

Figure 7.6
Radiographs of the (a) wrist and hand and (b) the ankle and foot. Several kinds of synovial joints (freely moveable joints) are contained in these regions.
1. Hinge joints—interphalangeal joints of the second digit (index finger)
2. Condyloid joint—metacarpophalangeal joint of the second digit (index finger)
3. Saddle joint—carpometacarpal joint of the pollex (thumb)
4. Gliding joint—intercarpal joint between capitate and scaphoid bones
5. Condyloid joint—radiocarpal joint
6. Condyloid joint at the ankle
7. Gliding joints between tarsal bones
8. Hinge joint between phalanges of digit

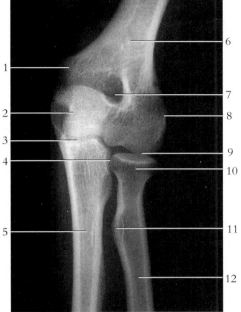

Figure 7.7
A radiograph of the elbow region depicting two types of synovial joints. The humeroulnar joint is a hinge join that permits movement along a single plane. The proximal radioulnar and humeroradial joints are both pivot joints that permit rotational movement.

1. Medial epicondyle
2. Olecranon
3. Humeroulnar joint
4. Proximal radioulnar joint
5. Ulna
6. Humerus
7. Olecranon fossa
8. Lateral epicondyle
9. Humeroradial joint
10. Head of radius
11. Radial tuberosity
12. Radius

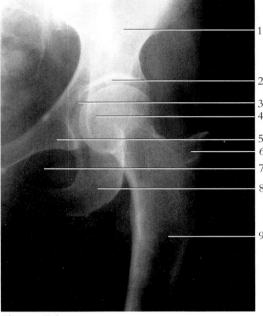

Figure 7.8
Tha ball-and-socket joint seen in this radiograph of the hip is a synovial joint. The coxal joint (hip joint) is formed by the head of the femur articulating with the acetabulum of the os coxae.

1. Ilium
2. Acetabulum
3. Ball-and-socket joint
4. Head of femur
5. Pubis
6. Greater trochanter
7. Obturator foramen
8. Ischium
9. Femur

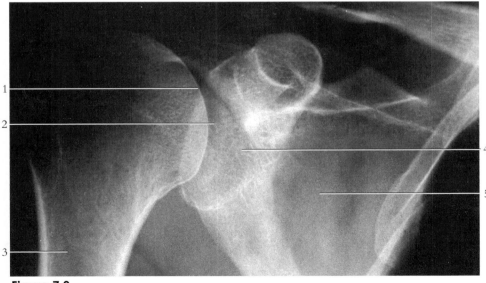

Figure 7.9
The ball-and-socket joint seen in this radiograph of the shoulder is a synovial joint. The glenohumeral joint (shoulder joint) is formed by the head of the humerus articulating with the glenoid fossa of the scapula.

1. Head of humerus
2. Ball-and-socket joint
3. Humerus
4. Glenoid fossa
5. Scapula

(a)
Flexion at left shoulder, elbow, and knee joints. Extension of right shoulder, elbow, and knee joints.

(b)
Maximal flexion at each of the principal body joints.

(c)
Rotation at the joints of the neck; elevation at the shoulder joint; and flexion at the right elbow and right wrist joints.

(d)
Abduction at the shoulder and hip joints. Also, abduction at the metacarpophalangeal joints at the base of the digits.

(e)
Adduction at the shoulder and hip joints. Also, adduction at the joints of the fingers and toes.

(f)
Rotation at joints of the vertebral column.

(g)
Lateral bending at joints of the vertebral column.

(h)
Flexion at the joints of the vertebral column.

(i)
Hyperextension at the joints of the vertebral column.

(j)
Flexion at the shoulder, hip, and knee joints on right side of body; extension at the elbow and wrist joints; plantar flexion at the right ankle joint.

(k)
Extension at the shoulder and hip joints on right side of body.

Figure 7.10
A photograghic summary of joint movements (a-k).

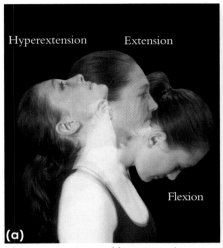

Flexion, extension, and hyperextension at the cervical intervertebral joints of the neck.

Adduction and abduction of the right arm at the shoulder joint and fingers at the metacarpophalangeal joints

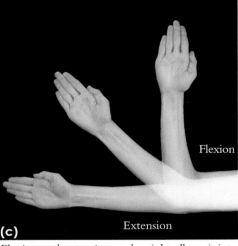

Flexion and extension at the right elbow joint.

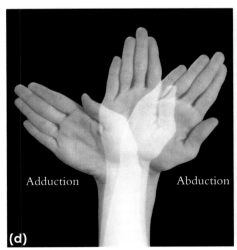

Abduction and adduction of the right hand at the wrist.

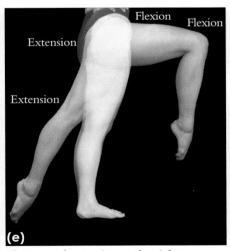

Flexion and extension at the right hip and knee joints.

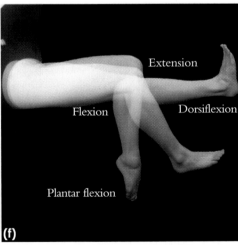

Flexion and extension at the knee joint and plantar flexion and dorsiflexion at the ankle joint.

Figure 7.11
A visualization of angular movements permitted at synovial joints (a–f).

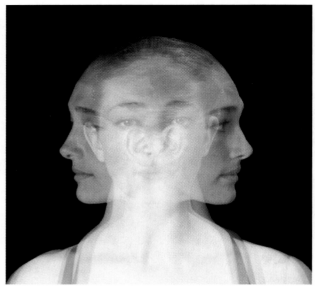

Figue 7.12
Rotation of the head at the cervical vertebrae-principally at the atlantoaxial joint.

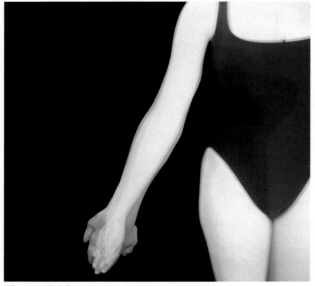

Figure 7.13
Rotation of the antebrachium (forearm) at the proximal radioulnar joint-an example of pronation and supination.

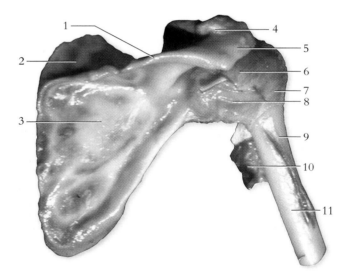

Figure 7.14
A posterior view of the right scapula, proximal humerus, and glenohumeral (shoulder) joint.
 1. Spine of scapula
 2. Supraspinous fossa
 3. Infraspinous fossa
 4. Coracoacrominal ligament
 5. Acromion
 6. Tendon of infraspinatus muscle
 7. Tendon of teres minor muscle
 8. Articular capsule of glenohumeral joint
 9. Origin of lateral head of triceps brachii muscle
10. Teres major muscle
11. Humerus

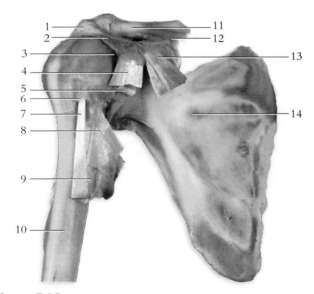

Figure 7.15
An anterior view of the right scapula, proximal humerus, and glenohumeral (shoulder) joint.
 1. Acromion
 2. Coracoid process of scapula
 3. Articular capsule of glenohumeral joint
 4. Tendon of short head of biceps brachii muscle
 5. Tendon of coracobrachialis muscle
 6. Tendon sheath
 7. Tendon of long head of biceps brachii muscle
 8. Tendon of latissimus dorsi muscle
 9. Tendon of teres major muscle
10. Humerus
11. Clavicle
12. Subclavius muscle
13. Tendon of pectoralis minor muscle
14. Subscapular fossa

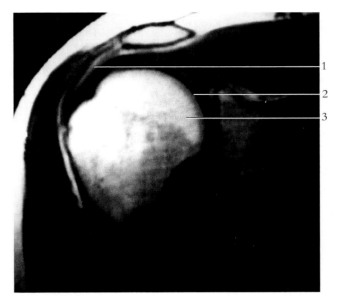

Figure 7.16
Structure of the humeral joint (shoulder joint) as
seen in a MR image.
1. Subdeltoid bursa
2. Articular cavity of shoulder joint
3. Head of humerus

Figure 7.17
An anterior view of the right elbow region.
1. Humerus
2. Lateral epicondyle of humerus
3. Articular capsule of elbow joint
4. Annular ligament
5. Head of radius
6. Radius
7. Interosseous ligament
8. Medial epicondyle
9. Ulnar collateral ligament
10. Ulna

Figure 7.18
A posterior view of the
right elbow region.
1. Olecranon fossa
2. Articular capsule of elbow joint
3. Lateral epicondyle
4. Olecranon
5. Ulna
6. Humerus
7. Medial epicondyle
8. Coronoid process of ulna
9. Head of radius
10. Radius

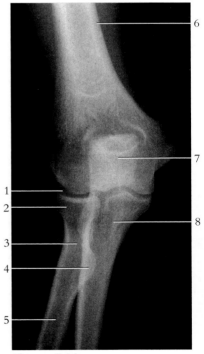

Figure 7.19
Structure of the elbow joint as
seen in a radiograph.
1. Articular cavity of elbow joint
2. Head of radius
3. Neck of radius
4. Radial tuberosity
5. Radius
6. Humerus
7. Olecranon
8. Ulna

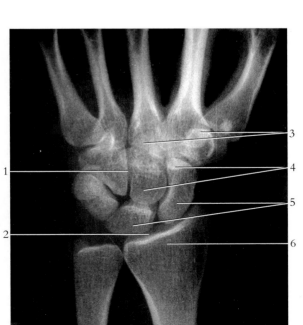

Figure 7.20
Structure of the wrist joint as
seen in a MR image.
1. Gliding joint
2. Condyloid joint
3. Heads of second and third metacarpal bones
4. Distal carpal bones (trapezoid and capitate bones)
5. Proximal carpal bones (scaphoid and lunate bones)
6. Distal epiphysis of radius

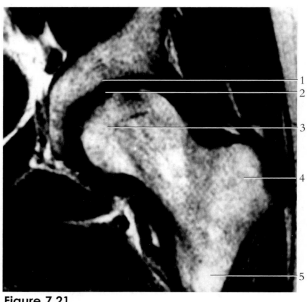

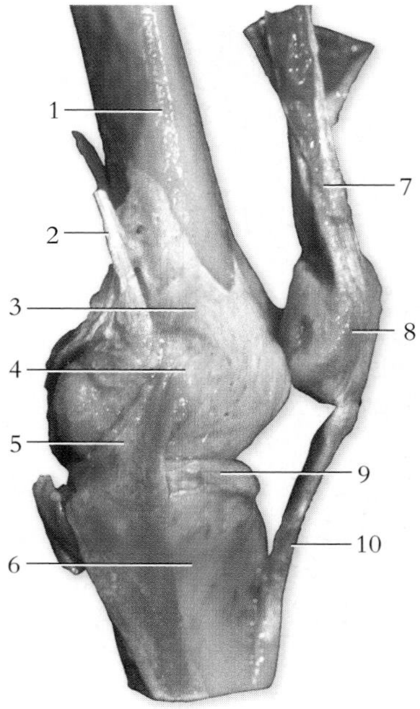

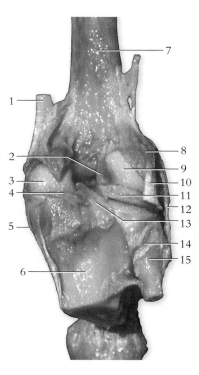

Figure 7.22
A medial view of the right tibiofemoral (knee) joint
1. Femur
2. Tendon of adductor magnus muscle
3. Articular capsule of tibiofemoral (knee) joint
4. Medial epicondyle of femur
5. Tibial collateral ligament
6. Tibia
7. Tendon of quadriceps femoris muscle
8. Patella
9. Articular cartilage of tibia
10. Patellar ligament

Figure 7.21
Structure of the coxal joint (hip joint) as seen in a MR image.
1. Acetabulum
2. Articular cavity of hip joint
3. Head of femur
4. Greater trochanter of femur
5. Body of femur

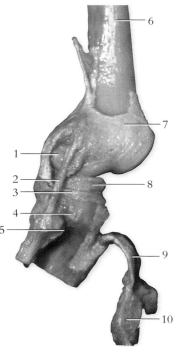

Figure 7.23
A lateral view of right tibiofemoral (knee) joint.
1. Bursa
2. Lateral collateral ligament
3. Articular cartilage
4. Proximal epiphysis
5. Tibiofibular joint
6. Femur
7. Articular capsule
8. Lateral meniscus
9. Patellar ligament
10. Patella

Figure 7.24
A posterior view of right tibiofemoral (knee) joint.
1. Tendon of adductor magnus muscle
2. Anterior cruciate ligament
3. Medial femoral condyle
4. Medial meniscus
5. Medial collateral ligament
6. Tibia
7. Femur
8. Lateral epicondyle of femur
9. Lateral femoral condyle
10. Tendon of popliteus muscle
11. Lateral meniscus
12. Lateral collateral ligament
13. Posterior cruciate ligament
14. Proximal tibiofibular joint
15. Head of fibula

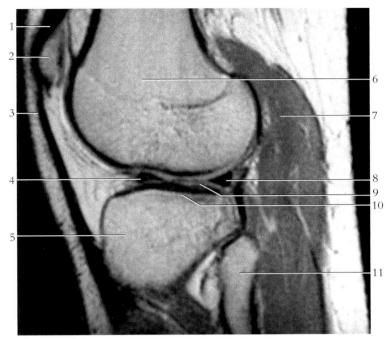

Figure 7.25
A sagittal MR image of tibiofemoral joint (knee joint).
1. Tendon of quadriceps muscle
2. Patella
3. Patellar ligament
4. Lateral meniscus, anterior horn
5. Head of tibia
6. Femur
7. Gastrocnemius muscle
8. Lateral meniscus, posterior horn
9. Synovial fluid
10. Articular cartilage
11. Head of fibula

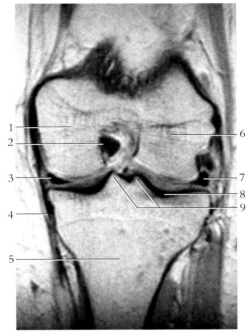

Figure 7.26
A posterior MR image of tibiofemoral joint (knee joint).
1. Medial condyle
2. Insertion of posterior cruciate ligament
3. Medial meniscus
4. Tibial collateral ligament
5. Tibia
6. Lateral condyle of femur
7. Lateral meniscus
8. Synovial fluid
9. Intercondylar eminences

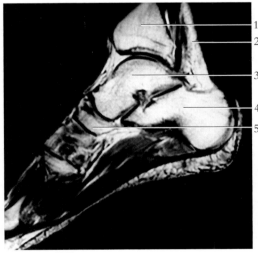

Figure 7.27
A sagittal MR image of the talocrual (ankle) and tarsal joints of the foot.
1. Tibia
2. Tendo calcaneus
3. Talus
4. Calcaneus
5. Navicular bone

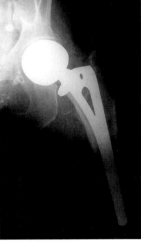

Figure 7.28
A radiograph of the hip showing a joint prosthesis.

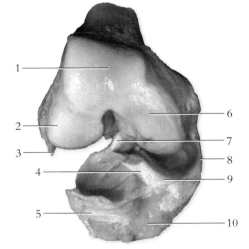

Figure 7.29
An anterior view of right tibiofemoral (knee) joint as it is flexed.
1. Patellar articular surface
2. Lateral femoral condyle
3. Lateral collateral ligament
4. Anterior cruciate ligament
5. Articular cartilage of tibia
6. Medial femoral condyle
7. Posterior cruciate ligament
8. Medial collateral ligament
9. Medial meniscus
10. Tibia

8 Muscular System

Most skeletal muscles span joints and are attached to a bone at both ends by a **tendon**. Certain tendons, especially at the wrist and ankle, are enclosed individually by **tendon sheaths**. A group of tendons in these areas is also covered by a **retinaculum**. The **origin** of a muscle is the more stationary attachment, and the **insertion** is the more moveable attachment (fig. 8.1). **Synergistic muscles** contract together. **Antagonistic muscles** perform in opposition to a synergistic group of muscles.

Fascia is a fibrous connective tissue that covers muscle and attaches to the skin. **Superficial fascia** secures the skin to the underlying muscle. **Deep fascia** is an inward extension of the superficial fascia and surrounds adjacent muscles compartmentalizing and binding them into functional groups.

A skeletal muscle fiber is an elongated, multinucleated, striated cell (fig. 8.2). Each fiber is surrounded by a cell membrane, called a **sarcolemma**, and the cytoplasm within the cell is called **sarcoplasm**. Each fiber contains many parallel, thread-like structures called **myofibrils**. Each myofibril is composed of smaller strands called **myofilaments** that contain the contractile proteins, **actin** and **myosin**. The regular spatial organization of the contractile proteins within the myofibrils forms the cross-banding striations characteristic of skeletal muscle. A network of membranous channels, called the **sarcoplasmic reticulum**, extends throughout the cytoplasm.

The myofibrils of a skeletal muscle fiber are arranged into compartments called **sarcomeres**. The dark and light striations of the sarcomere are due to the arrangement of the thick (myosin) and thin (actin) filaments. The dark bands are called **A–bands** and the lighter bands, containing only actin filaments, are called **I–bands**. The outer regions of the A bands contain actin and myosin; however, the lighter central regions **(H–zones)** of the A bands contain only myosin. The I bands are bisected by dark **Z–lines** where the actin filaments of adjacent sarcomeres join. Muscle fiber contraction results from the interaction of contractile proteins (actin and myosin myofilaments) in which the length of the sarcomeres is reduced.

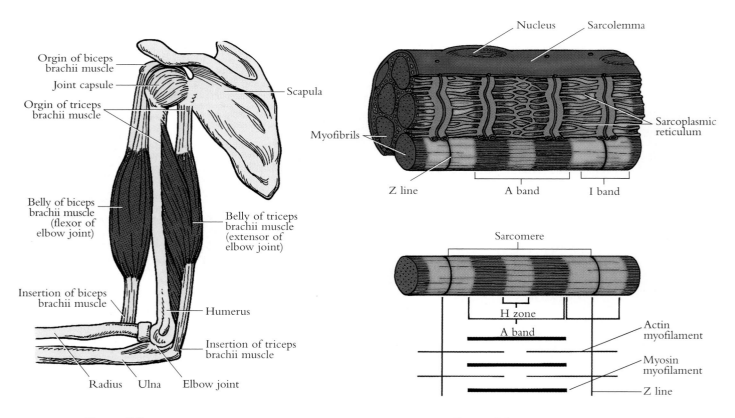

Figure 8.1
Skeletomuscular relationship.

Figure 8.2
Structure of a skeletal muscle fiber.

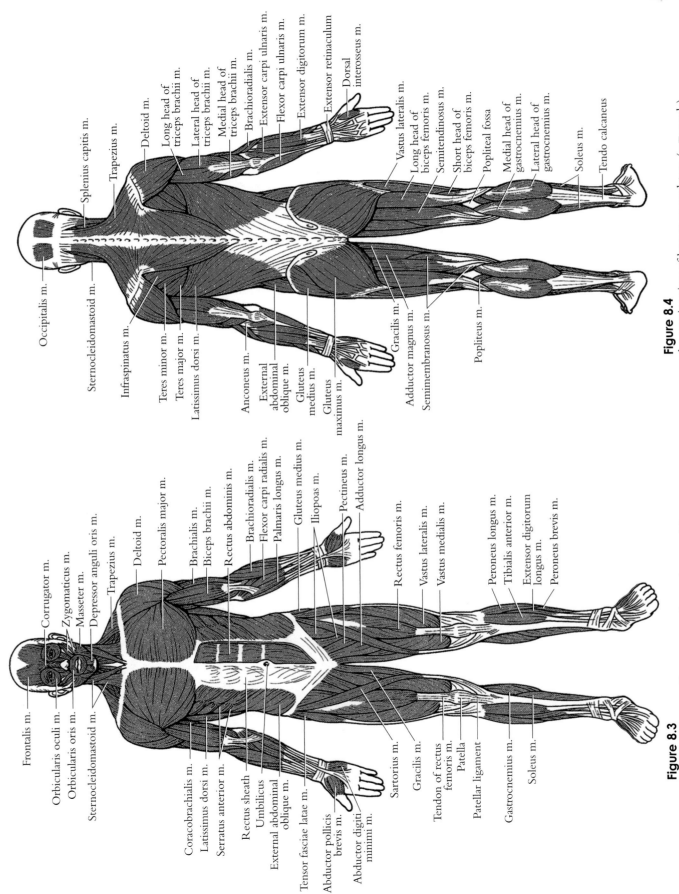

Figure 8.4
A posterior view of human musculature (m=muscle).

Figure 8.3
An anterior view of human musculature (m=muscle).

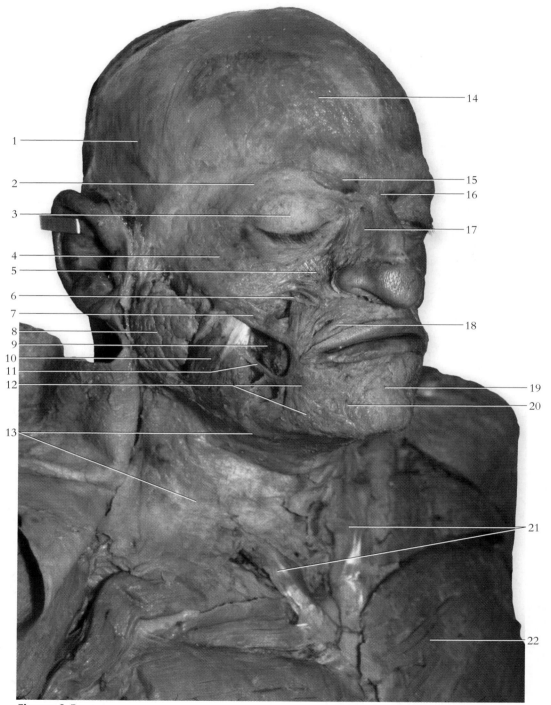

Figure 8.5

Muscles of the anterior head and neck regions.

1. Temporalis m.
2. Orbicularis oculi m.
3. Palpebral part of orbicularis oculi m.
4. Orbital part of orbicularis oculi m.
5. Levator labii superioris m.
6. Zygomaticus minor m.
7. Zygomaticus major m.
8. Parotid gland
9. Buccinator m.
10. Masseter m.
11. Risorius m.
12. Depressor anguli oris m.
13. Platysma m.
14. Frontalis m.
15. Corrugator supercilii m.
16. Procerus m.
17. Nasalis m.
18. Orbicularis oris m.
19. Mentalis m.
20. Depressor labii inferioris m.
21. Sternocleidomastoid m.
22. Pectoralis major m.

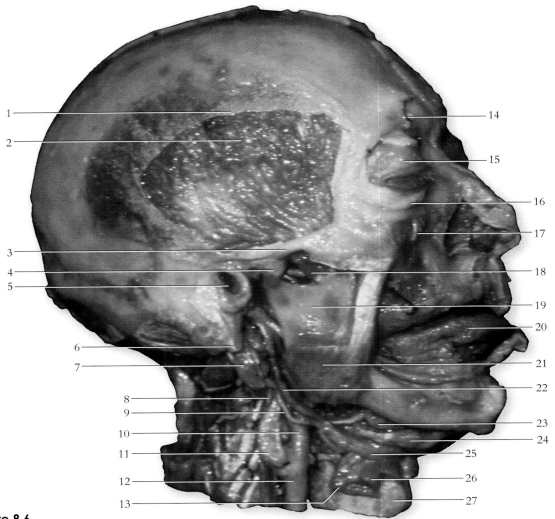

Figure 8.6
A lateral view of the deep structures of the head and neck.

1. Temporal fascia
2. Temporalis muscle
3. Zygomatic arch
4. Joint capsule of temporomandibular joint
5. External acoustic canal
6. Mastoid process of temporal bone
7. Posterior belly of digastric muscle
8. Vagus nerve
9. Hypoglossal nerve
10. Internal cartoid artery
11. External cartoid artery
12. Common carotid artery
13. Omohyoid muscle
14. Supraorbital nerve
15. Superior tarsal plate
16. Palpebral fascia
17. Infraorbital nerve
18. Lateral pterygoid muscle
19. Mandible
20. Tongue
21. Masseter muscle
22. Stylohyoid muscle
23. Mandibular gland
24. Anterior belly of digastric muscle
25. Mylohyoid muscle
26. Sternohyoid muscle
27. Thyroid cartilage of larynx

Table 8.1 Muscles of mastication.

Chewing Muscle	Origin(s)	Insertion(s)	Action	Innervation
Temporalis	Temporal fossa	Coronoid process of mandible	Elevates mandible	Trigeminal n.
Masseter	Zygomatic arch	Lateral ramus of mandible	Elevates mandible	Trigeminal n.
Medial pterygoid	Sphenoid bone	Medial ramus of mandible	Depresses and laterally moves mandible	Trigeminal n.
Lateral pterygoid	Sphenoid bone	Anterior side of condylar process of mandible	Protracts mandible	Trigeminal n.

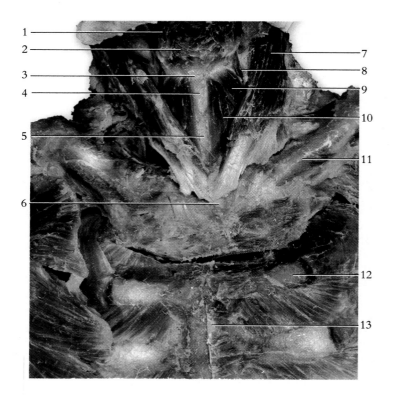

Figure 8.7
An anterior view of the neck muscles.
1. Digastric m.
2. Mylohyoid m.
3. Hyoid bone
4. Thyroid cartilage
5. Trachea
6. Manubrium of sternum
7. Sternocleidomastoid m.
8. Thyrohyoid m.
9. Omohyoid m. (superior belly)
10. Sternohyoid m.
11. Clavicle
12. Rib
13. Body of sternum

Table 8.2 Muscles of the neck.

Neck Muscle	Origin(s)	Insertion(s)	Action	Innervation
Sternocleidomastoid	Sternum and clavicle	Mastoid process of temporal bone	Flexes neck; rotates head to side	Accessory n.
Digastric	Inferior border of mandible and mastoid process of temporal bone	Hyoid bone	Opens mouth; elevates hyoid bone	Trigeminal n. (ant. belly) facial n. (post. belly)
Mylohyoid	Inferior border of mandible	Hyoid bone and median raphe	Elevates hyoid bone and floor of mouth	Trigeminal n.
Geniohyoid	Medial surface of mandible at chin	Hyoid bone	Elevates hyoid bone	Spinal n. (C1)
Stylohyoid	Styloid process of temporal bone	Hyoid bone	Elevates and retracts tongue	Facial n.
Sternohyoid	Manubrium	Hyoid bone	Depresses hyoid bone	Spinal n. (C1-C3)
Sternothyroid	Manubrium	Thyroid cartilage	Depresses thyroid cartilage	Spinal n. (C1-C3)
Thyrohyoid	Thyroid cartilage	Hyoid bone	Depresses hyoid bone; elevates thyroid cartilage	Spinal n. (C1-C3)
Omohyoid	Superior border of scapula	Hyoid bone	Depresses hyoid bone	Spinal n. (C1-C3)

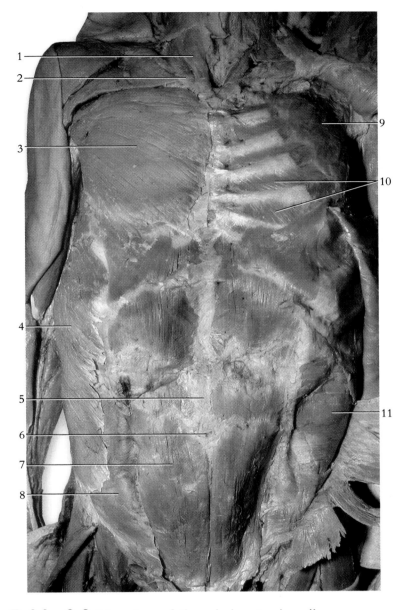

Figure 8.8
An anterolateral view of the trunk.
  1. Sternocleidomastoid m.
  2. Clavicle
  3. Pectoralis major m.
  4. External abdominal oblique m.
  5. Linea alba
  6. Umbilicus
  7. Rectus abdominis m.
  8. Internal abdominal oblique m.
  9. Pectoralis minor m.
 10. External intercostal mm.
 11. Transverse abdominis m.

Table 8.3 Muscles of the abdominal wall.

Abdominal Muscle	Origin(s)	Insertion(s)	Action	Innervation
External abdominal oblique	Lower eight ribs	Iliac crest and linea alba	Compresses abdomen; lateral rotation	Intercostal nn. Iliohypogastric n. Ilioinguinal n.
Internal abdominal oblique	Iliac crest, lumbodorsal fascia, inguinal ligament	Linea alba and costal cartilages of last three of four ribs	Compresses abdomen; lateral rotation	Intercostal nn. Iliohypogastric n. Ilioinguinal n.
Transversus abdominis	Iliac crest, lumbodorsal fascia, inguinal ligament, costal cartilages of last six ribs	Xiphoid process, linea alba, pubis	Compresses abdomen;	Intercostal nn. Iliohypogastric n. Ilioinguinal n.
Rectus abdominis	Pubic crest and symphysis pubis	Xiphoid process and costal cartilages of fifth to seventh ribs	Flexes vertebral column	Intercostal nn.

Figure 8.9

An anterolateral view of the trunk.

1. Deltoid m.
2. Pectoralis major m.
3. Biceps brachii m. (long head)
4. Brachialis m.
5. Serratus anterior m.
6. Brachioradialis m.
7. External abdominal oblique m.
8. Umbilicus
9. Tendon of sternocleidomastoid m.
10. Sternum.
11. Xiphoid process
12. Tendinous inscriptions of rectus abdominis m.
13. Rectus abdominis m.

Figure 8.10

A posterolateral view of the trunk.

1. Trapezius m.
2. Triangle of ausculation
3. Latissimus dorsi mm.
4. Vertebral column (spinous processes)
5. Infraspinatus m.
6. Deltoid m.
7. Teres minor m.
8. Teres major m.
9. Serratus anterior mm.
10. Rib
11. External abdominal oblique m.
12. Iliac crest
13. Gluteus medius m.
14. Gluteus maximus m.

Table 8.4 Muscles that act on the pectoral girdle.

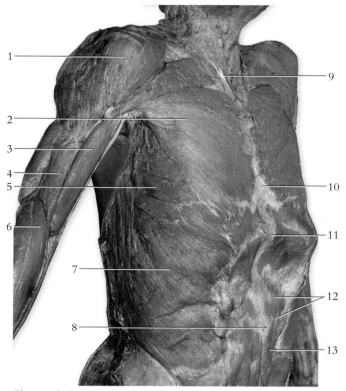

Pectoral Muscle	Origin(s)	Insertion(s)	Action	Innervation
Serratus anterior	Upper eight or nine ribs	Anterior medial border of scapula	Pulls scapula forward and upward	Long thoracic n.
Pectoralis minor	Sternal ends of third, fourth,and fifth ribs	Coracoid process of scapula	Pulls scapula forward and downward	Medial and lateral pectoral nn.
Subclavius	First rib	Subclavian groove of clavical	Draws clavicle downward	Spinal nn. C5, C6
Trapezius	Occipital bone and spines of cervical and thoracic vertebrae	Clavicle, acromion and spineof scapula	Elevates, depresses, and adducts scapula; hyperextends neck; braces shoulder	Accessory n
Levator scapulae	First to fourth cervical vertebrae	Superior border of scapula	Elevates scapula	Dorsal scapular n.
Rhomboideus major	Spines of second to fifth thoracic vertebrae	Medial border of scapula	Elevates and adducts scapula	Dorsal scapular n.
Rhomboideus minor	Seventh cervical and first thoracic vertebrae	Medial border of scapula	Elevates and adducts scapula	Dorsal scapular n.

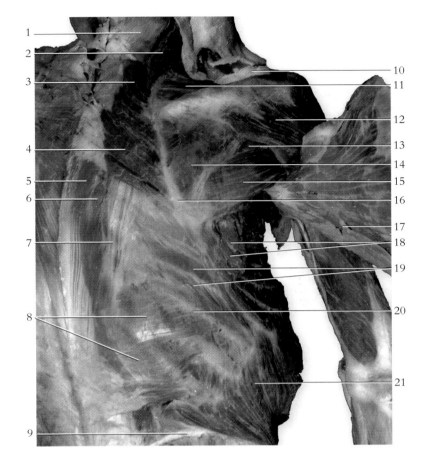

Figure 8.11
The deep muscles of the back and right side.
1. Splenius capitis m.
2. Levator scapulae m.
3. z   m.
4. Rhomboideus major m.
5. Spinalis thoracis m.
6. Longissimus thoracis m.
7. Iliocostalis thoracis m.
8. Serratus posterior inferior m.
9. Iliac crest
10. Spine of the scapula
11. Supraspinatus m.
12. Deltoid m.
13. Teres minor m.
14. Infraspinatus m.
15. Teres major m.
16. Scapula (medial border)
17. Latissimus dorsi m. (reflected)
18. Serratus anterior m.
19. External intercostal mm.
20. Rib
21. External abdominal oblique m.

Table 8.5 Muscles of the vertebral column.

Spinal Muscle	Origin(s)	Insertion(s)	Action	Innervation
Quadratus lumborum	Iliac crest and lower three lumbar vertebrae	Twelfth rib and upper four lumbar vertebrae	Extends lumbar region; flexes vertebral column laterally	Intercostal n. T12 and Lumbar nn. L2-L4
Erector spinae				
Iliocostalis lumborum	Crest of ilium	Lower six ribs	Extends lumbar region	Post. rami of lumbar nn.
Iliocostalis thoracis	Lower six ribs	Upper six ribs	Extends thoracic region	Post. rami of thoracic nn.
Iliocostalis cervicis	Angles of third–sixth ribs	Transverse processes of fourth–sixth cervival vertebrae	Extends cervical region	Post. rami of cervical nn.
Longissimus thoracis	Transverse processes of lumbar vertebrae	Transverse processes of all the thoracic vertebrae and lower nine ribs	Extends thoracic region	Post. rami of spinal nn.
Longissimus cervicis	Transverse processes of upper four or five thoracic vertebrae	Transverse processes of second–sixth cervical vertebrae	Extends cervical region and lateral flexion	Post. rami of spinal nn.
Longissimus capitis	Transverse processes of upper five thoracic vertebrae	Posterior margin of cranium and mastoid process of temporal bone	Extends head; acting separately, turns face toward that side	Post. rami of middle and lower cervical nn.
Spinalis thoracis	Spinous processes of upper lumbar and lower thoracic vertebrae	Spinous processes of upper thoracic vertebrae	Extends vertebral column	Post. rami of spinal nn.

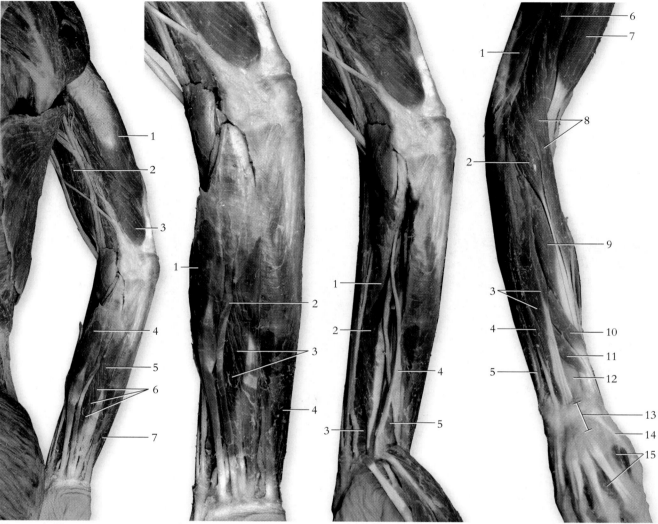

Figure 8.12
The medial brachium and superficial flexors of the right antebrachium.
1. Triceps brachii m. (lateral head)
2. Biceps brachii m. (short head)
3. Triceps brachii m. (medial head)
4. Flexor carpi radialis m.
5. Palmaris longus m.
6. Superficial digital flexor m.
7. Flexor carpi ulnaris m.

Figure 8.13
An anterior view of the superficial muscles of the right forearm.
1. Flexor carpi radialis m.
2. Palmaris longus m.
3. Superficial digital flexor m.
4. Flexor carpi ulnaris m.

Figure 8.14
An anterior view of the deep muscles of the right forearm.
1. Pronator teres m.
2. Flexor pollicis longus m.
3. Pronator quadratus m.
4. Median nerve
5. Deep digital flexor m.

Figure 8.15
A posterior view of the superficial muscles of the right forearm.
1. Triceps brachii m. (medial head)
2. Extensor carpi radialis longus m.
3. Extensor digitorum m.
4. Extensor carpi minimi m.
5. Extensor carpi ulnaris m.
6. Brachialis m.
7. Biceps brachii m. (long head)
8. Brachioradialis m.
9. Extensor carpi radialis brevis m.
10. Abductor pollicis longus m.
11. Extensor pollicis brevis m.
12. Radius
13. Extensor retinaculum
14. Tendon of extensor pollicis longus m.
15. Dorsal interosseous mm.

Table 8.6 Muscles that act on the brachium (upper arm).

Axial or Scapular Muscle	Origin(s)	Insertion(s)	Action	Innervation
Pectoralis major	Clavicle, sternum, costal cartilages of second to sixth ribs	Greater tubercle of humerus	Flexes, adducts, and rotates shoulder joint medially	Medial and lateral pectoral nn.
Latissimus dorsi	Spines of sacral, lumbar, and lower thoracic vertebrae; lower ribs	Intertubercular groove of humerus	Extends, adducts, and rotates shoulder joint medially; adducts shoulder joint	Thoracodorsal n.
Deltoid	Clavicle, acromion and spine of scapula	Deltoid tuberosity of humerus	Abducts, extends, or flexes shoulder joint	Axillary n.
Supraspinatus	Supraspinous fossa of scapula	Greater tubercle of humerus	Abducts and laterally rotates shoulder joint	Suprascapular n.
Infraspinatus	Infraspinous fossa of scapula	Greater tubercle of humerus	Rotates shoulder joint laterally	Suprascapular n.
Teres major	Inferior angle and lateral border of scapula	Intertubercular groove of humerus	Extends, adducts, and rotates shoulder joint medially	Lower subscapular n.
Teres minor	Lateral border of scapula	Greater tubercle of humerus cartilage	Rotates shoulder joint laterally	Axillary n.
Subscapularis	Subscapular fossa	Subscapular fossa	Rotates shoulder joint medially	Subscapular n.
Coracobrachialis	Coracoid process of scapula	Body of humerus	Flexes and adducts shoulder joint	Musculocutaneous n.

Table 8.7 Muscles that act on the antebrachium (forearm).

Brachial Muscle	Origin(s)	Insertion(s)	Action	Innervation
Biceps brachii	Coracoid process and tuberosity above glenoid	Radial tuberosity	Flexes elbow joint; supinates forearm and hand at radioulnar joint	Musculocutaneous n.
Brachialis	Anterior body of humerus	Coronoid process of ulna	Flexes elbow joint	Musculocutaneous n.
Brachioradialis	Lateral supracondylar ridge of humerus	Proximal to styloid process of radius	Flexes elbow joint	Radial n.
Triceps brachii	Tuberosity below glenoid fossa: lateral and medial surfaces of humerus	Olecranon of ulna	Extends elbow joint	Radial n.
Anconeus	Lateral epicondyle of humerus	Olecranon of ulna	Extends elbow joint	Radial n.

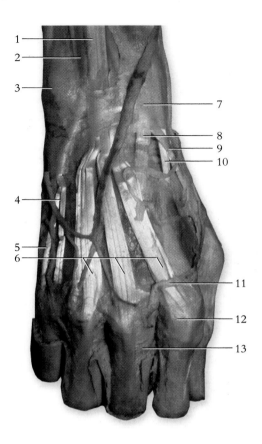

Figure 8.16
Posterior view of right hand showing superficial structures.
1. Tendon of extensor digitorum communis m.
2. Extensor digiti minimi m.
3. Styloid process of ulna
4. Tendon of extensor carpi ulnaris m.
5. Tendon of extensor digiti minimi m.
6. Tendons of extensor digitorum communis mm.
7. Extensor retinaculum
8. Tendon of extensor carpi radialis longus m.
9. Tendon of extensor carpi radialis brevis m.
10. Tendon of extensor pollicis longus m.
11. Superficial venous arch
12. Articular capsule of metacarpophalangeal joint
13. Fibrous digital sheath

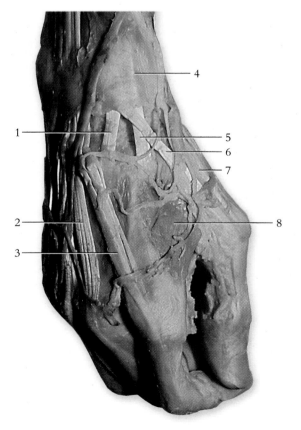

Figure 8.17
Lateral view of wrist and hand.
1. Tendon of extensor carpi radialis longus m.
2. Tendon of extensor digitorum communis m.
3. Tendon of extensor indicis m.
4. Extensor retinaculum
5. Tendon of extensor carpi radialis brevis m.
6. Tendon of extensor pollicis longus m.
7. Tendon of extensor pollicis brevis m.
8. Dorsal interosseous m.

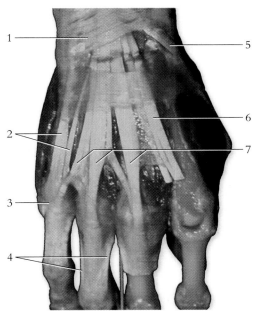

Figure 8.18
A posterior view of hand showing extensor tendons.
1. Extensor retinaculum
2. Tendon of extensor digiti minimi m.
3. Articular capsule of metacarpophangeal joint
4. Extensor expansions
5. Tendon of extensor pollicis longus m.
6. Tendon of extensor indicis m. (cut)
7. Tendons of extensor digitorum m.

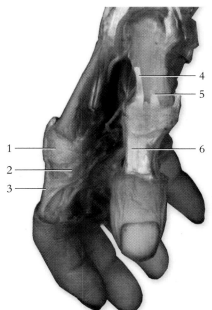

Figure 8.19
A lateral view of hand showing deep tendons.
1. Articular capsule of metacarpophangeal joint
2. Tendon of lumbrical m.
3. Extensor expansion
4. Tendon of extensor pollicis longus m.
5. Tendon of extensor pollicis brevis m.
6. Extensor expansion of pollicis (thumb)

Figure 8.20
An anterior view of right hand
1. Pronator quadratus m.
2. Radial artery
3. Flexor synovial sheath
4. Opponens pollicis m.
5. Flexor pollicis brevis m.
6. Tendon of extensor pollicis brevis m.
7. Adductor pollicis mm.
8. Tendon of flexor pollicis longus m.
9. Ulnar nerve
10. Ulnar artery
11. Tendon of flexor carpi ulnaris m.
12. Opponens digiti minimi m.
13. Abductor for digiti minimi and flexor digiti minimi mm.
14. Lumbrical mm.
15. Tendons of deep digital flexor m. (cut)

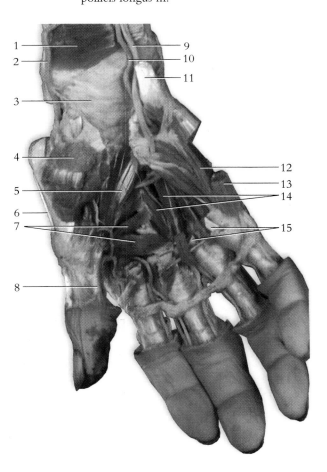

Table 8.8 Muscles that act on the wrist, hand, and fingers

Antebrachial Muscle	Origin(s)	Insertion(s)	Action	Innervation
Supinator	Lateral epicondyle of humerus and crest of ulna	Lateral surface of radius	Supinates forearm and hand	Radial n.
Pronator teres	Medial epicondyle of humerus	Lateral surface of radius	Pronates forearm and hand	Median n.
Pronator quadratus	Distal fourth of ulna	Distal fourth of radius	Pronates forearm and hand	Median n.
Flexor carpi radialis	Medial epicondyle of humerus	Base of second and third metacarpal bones	Flexes and abducts hand at wrist	Median n.
Palmaris longus	Medial epicondyle of humerus	Palmar aponeurosis	Flexes wrist	Median n.
Flexor carpi ulnaris	Medial epicondyle of humerus and olecranon of ulna	Carpal and metacarpal bones	Flexes and adducts wrist	Ulnar n.
Superficial digital flexor	Medial epicondyle of humerus and coronoid process of ulna	Middle phalanges of digits II–V	Flexes wrist and digits	Median n.
Deep digital flexor	Proximal two-thirds of ulna and interosseous ligament	Distal phalanges of digits II–V	Flexes wrist and digits	Median and ulnar n.
Flexor pollicis longus	Body of radius and coronoid process of ulna	Distal phalanx of thumb	Flexes joints of thumb	Median n.
Extensor carpi radialis longus	Lateral epicondyle of humerus	Second metacarpal bone	Extends and abducts wrist	Radial n.
Extensor carpi radialis brevis	Lateral epicondyle of humerus	Third metacarpal bone	Extends and abducts wrist	Radial n.
Extensor digitorum communis	Lateral epicondyle of humerus	Posterior surfaces of digits II–V	Extends wrist and phalanges	Radial n.
Extensor digiti minimi	Lateral epicondyle of humerus	Extensor aponeurosis of fifth digit	Extends joints of fifth digit and wrist	Radial n.
Extensor carpi ulnaris	Lateral epicondyle of humerus and olecranon of ulna	Base of fifth metacarpal bone	Extends and adducts wrist	Radial n.
Extensor pollicis longus	Lateromedial body of ulna	Base of distal phalanx of thumb	Extends joints of thumb; abducts joints of hand	Radial n.
Extensor pollicis brevis	Distal body of radius and interosseous ligament	Base of first phalanx of thumb	Extends joints of thumb; abducts joints of hand	Radial n.
Abductor pollicis longus	Distal radius and ulna and interosseous ligament	Base of first metacarpal bone	Abducts joints of thumb and joints of hand	Radial n.

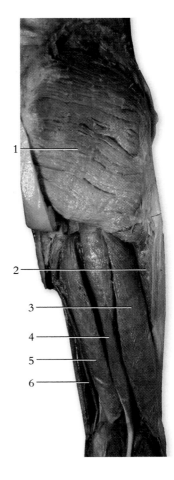

Figure 8.21
The superficial muscles of gluteal and thigh regions.
1. Gluteus maximus m.
2. Vastus lateralis m.
3. Biceps femoris m.
4. Semitendinosus m.
5. Semimembranosus m.
6. Gracilis m.

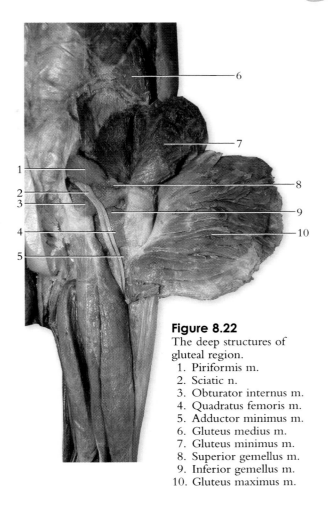

Figure 8.22
The deep structures of gluteal region.
1. Piriformis m.
2. Sciatic n.
3. Obturator internus m.
4. Quadratus femoris m.
5. Adductor minimus m.
6. Gluteus medius m.
7. Gluteus minimus m.
8. Superior gemellus m.
9. Inferior gemellus m.
10. Gluteus maximus m.

Table 8.9 Muscles that act on the thigh at the hip joint.

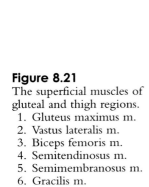

Pelvic Muscle	Origin(s)	Insertion(s)	Action	Innervation
Iliacus	Iliac fossa	Lesser trochanter of femur, along with psoas major	Flexes and rotates thigh laterally at the hip joint; flexes joints of vertebral column	Femoral n.
Psoas major	Transverse process of lumbar vertebrae	Lesser trochanter of femur, along with iliacus	Flexes and rotates thigh laterally at the hip joint; flexes joints of vertebral column	Spinal nn. L2, L3
Gluteus maximus	Iliac crest, sacrum, coccyx, aponeurosis of lumbar region	Gluteal tuberosity and iliotibial tract	Extends and rotates thigh laterally at the hip joint	Inferior gluteal n.
Gluteus medius	Lateral surface of ilium	Greater trochanter of femur	Abducts and rotates thigh medially at the hip joint	Superior gluteal n.
Gluteus minimus	Lateral surface of lowerhalf of ilium	Greater trochanter of femur	Abducts and rotates thigh medially at the hip joint	Superior gluteal n.
Tensor fasciae latae	Anterior border of ilium and iliac crest	Iliotibial tract	Abducts thigh at the hip joint	Superior gluteal n.

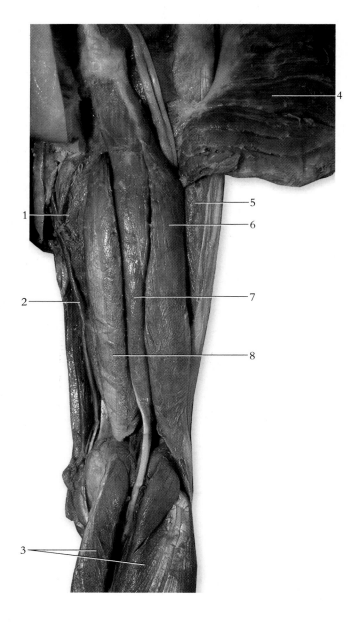

Figure 8.23
The deep posterior structures of thigh and popliteal regions.
1. Adductor magnus m.
2. Gracilis m.
3. Gastrocnemius m.
4. Gluteus maximus m.
5. Vastus lateralis m.
6. Biceps femoris m.
7. Semitendinosus m.
8. Semimembranosus m.

Table 8.10 Medial muscles that move the thigh at the hip joint.

Adductor Muscle	Origin(s)	Insertion(s)	Action	Innervation
Gracilis	Inferior edge of symphysis pubis	Proximomedial surface of tibia	Adducts thigh at hip joint; flexes and medially rotates leg at knee joint	Obturator n.
Pectineus	Pectineal line of pubis	Distal to lesser trochanterof femur	Adducts and flexes thigh at hip joint	Femoral n.
Adductor longus	Pubis—below pubic crest	Linea aspera of femur	Adducts, flexes, and laterally rotates thigh at hip joint	Obturator n.
Adductor brevis	Inferior ramus of pubis	Linea aspera of femur	Adducts, flexes, and laterally rotates thigh at hip joint	Obturator n.
Adductor magnus	Inferior ramus of ischium and inferior ramus of pubis	Linea aspera and medial epicondyle of femur	Adducts, flexes, and laterally rotates thigh at hip joint	Obturator and tibial nn.

Figure 8.24
An anterior view of the
right superior thigh
 1. Inguinal ligament
 2. Lateral femoral
    cutaneous nerve
 3. Superficial circumflex
    iliac artery
 4. Iliacus m.
 5. Femoral nerve
 6. Femoral artery
 7. Tensor fasciae latae m.
 8. Sartorius m.
 9. Rectus femoris m.
10. Femoral ring
11. Femoral vein
12. Pectineus m.
13. Great saphenous vein
14. Adductor longus m.

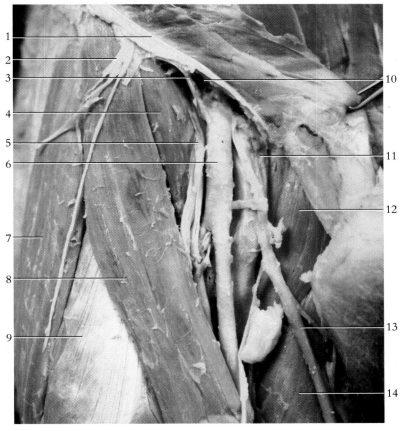

Table 8.11 Muscles of the thigh that act on the leg.

Thigh Muscle	Origin(s)	Insertion(s)	Action	Innervation
Sartorius	Anterior superior iliac spine	Medial surface of tibia	Flexes leg and thigh; abducts rotates thigh laterally; and rotates leg medially at hip joint	Femoral n.
Quadriceps femoris		Patella by common tendon, which continues as patellar ligament to tibial tuberosity	Extends leg at knee joint	Femoral n.
Rectus femoris	Anterior inferior iliac spine			
Vastus lateralis	Greater trochanter and linea aspera of femur			
Vastus medialis	Medial surface and linea aspera of femur			
Vastus intermedius	Anterior and lateral surfaces of femur			
Biceps femoris	Long head—ischial tuberosity; short head—linea aspera of femur	Head of fibula and proximolateral part of tibia	Flexes leg at knee joint; extends and laterally rotates thigh at hip joint	Tibial n.
Semitendinosus	Ischial tuberosity	Proximomedial surface of tibia	Flexes leg at knee joint; extends and medially rotates thigh at hip joint	Tibial n.
Semimembranosus	Ischial tuberosity	Proximomedial part of tibia	Flexes leg at knee joint; extends and medially rotates thigh at hip joint	Tibial n.

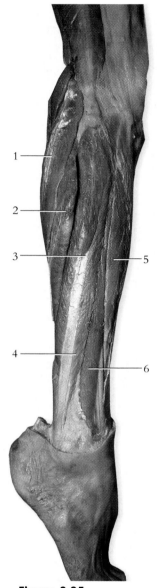

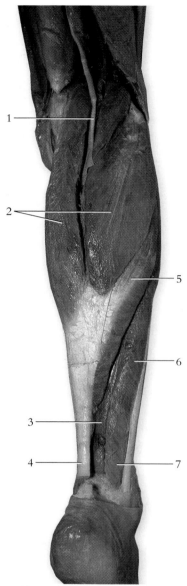

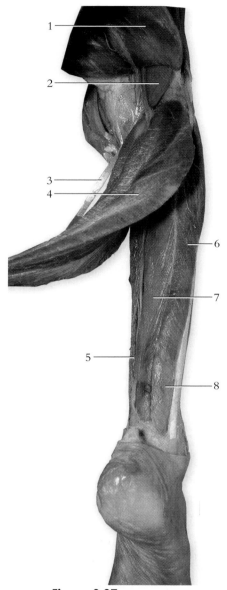

Figure 8.25
A lateral view of leg.
1. Gastrocnemius m.
2. Soleus m.
3. Peroneus longus m.
4. Peroneus brevis m.
5. Tibialis anterior m.
6. Peroneus tertius m.

Figure 8.26
A posterior superficial view of leg.
1. Tendon of semitendonosus m.
2. Gastrocnemius m.
3. Flexor hallucis longus m.
4. Tendo calcaneous (achilles tendon)
5. Soleus m.
6. Peroneus longus m.
7. Peroneus brevis m.

Figure 8.27
A posterior deep view of leg
1. Gastrocnemius m. (cut and
   reflected)
2. Plantaris m.
3. Tendon of plantaris m.
4. Soleus m. (cut and reflected)
5. Flexor digitorum longus m.
6. Peroneus longus m.
7. Flexor hallucis longus m.
8. Peroneus brevis m.

Table 8.12 Muscles of the leg that move the ankle, foot, and toes.

Leg Muscle	Origin(s)	Insertion(s)	Action	Innervation
Tibialis anterior	Lateral condyle and body of tibia	First metatarsal bone and first cuneiform bone	Dorsiflexes ankle; inverts foot and ankle	Deep fibular n.
Extensor digitorum longus	Lateral condyle of tibia and anterior surface of fibula	Extensor expansions of digits II–V	Extends digits II–V; dorsiflexes foot at ankle	Deep fibular n.
Extensor hallucis longus	Anterior surface of fibula and interosseous ligament	Distal phalanx of digit I	Extends joints of big toe; assists dorsiflexion of foot at ankle	Deep fibular n.
Peroneus tertius	Anterior surface of fibula and interosseous ligament	Dorsal surface of fifth metatarsal bone	Dorsiflexes and everts foot at ankle	Deep fibular n.
Peroneus longus	Lateral condyle of tibia and head and shaft of fibula	First cuneiform and metatarsal bone I	Plantar flexes and everts foot at ankle	Superficial fibular n.
Peroneus brevis	Lower aspect of fibula	Metatarsal bone V	Plantar flexes and everts foot at ankle	Superficial fibular n.
Gastrocnemius	Lateral and medial condyle of femur	Posterior surface of calcaneous	Plantar flexes foot at ankle; flexes knee joint	Tibial n.
Soleus	Posterior aspect of fibula and tibia	Calcaneous	Plantar flexes foot at ankle	Tibial n.
Plantaris	Supracondylar ridge of femur	Calcaneous	Plantar flexes foot at ankle	Tibial n.
Popliteus	Lateral condyle of femur	Upper posterior aspect of tibia	Flexes and medially rotates leg at knee joint	Tibial n.
Flexor hallucis longus	Posterior aspect of fibula	Distal phalanx of big toe	Flexes joint of distal phalanxof big toe	Tibial n.
Flexor digitorum longus	Posterior surface of tibia	Distal phalanges of digits II–V	Flexes joints of distal phalanges of digits II–V	Tibial n.
Tibialis posterior	Tibia and fibula and interosseous ligament bones II–IV	Navicular, cuneiform, cuboid, and metatarsal supports arches of foot	Plantar flexes and inverts foot at ankle;	Tibial n.

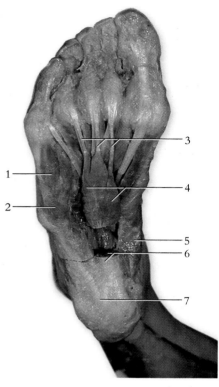

Figure 8.28
A superficial view of plantar region.
1. Flexor digiti minimi brevis m.
2. Abductor digiti minimi m.
3. Tendons of flexor digitorum brevis m.
4. Flexor digitorum brevis m.
5. Abductor hallucis m.
6. Plantar aponeurosis (cut)
7. Calcaneal tuberosity

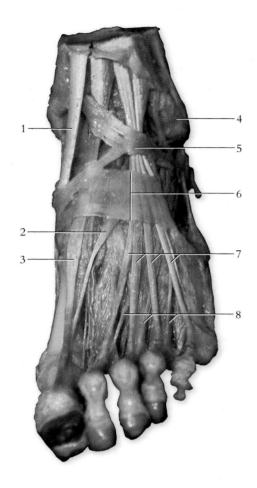

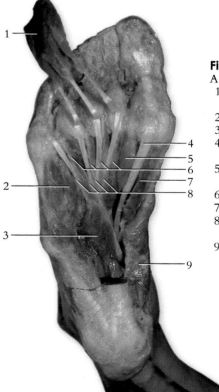

Figure 8.29
A deep view of plantar region.
1. Flexor digitorum brevis m. (reflected)
2. Interosseous m.
3. Quadratus plantae m.
4. Tendon of flexor hallucis longus m.
5. Oblique head of adductor hallucis m.
6. Lumbricals
7. Flexor hallucis brevis m.
8. Tendons of flexor digitorum longus m.
9. Abductor hallucis m.

Figure 8.30
An anterior view of dorsum of foot.
1. Tendon of tibialis anterior m.
2. Tendon of extensor hallucis brevis m.
3. Tendon of extensor hallucis longus m.
4. Lateral malleolus
5. Superior extensor retinaculum
6. Inferior extensor retinaculum
7. Tendon of extensor digitorum longus m.
8. Tendon of extensor digitorum brevis m.

Figure 8.31
A medial view of the right foot
 1. Tendon of tibialis anterior m.
 2. Extensor retinaculum
 3. Medial cuneiform
 4. Tendon of extensor digitorum longus m.
 5. First metatarsal bone
 6. Proximal phalanx of hallux
 7. Medial malleolus of tibia
 8. Tendo calcaneus
 9. Tendon of tibialis posterior m.
10. Tendon of flexor digitorum longus m.
11. Abductor hallucis m.

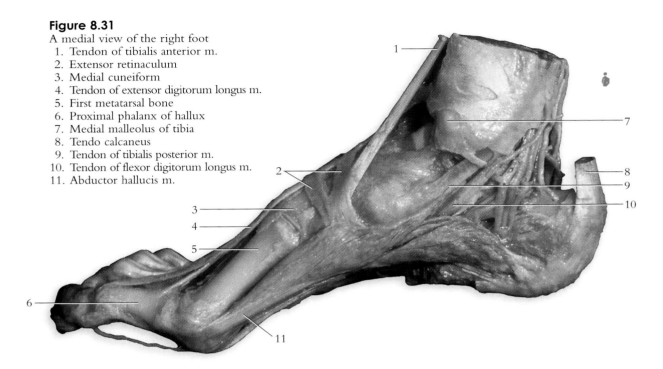

Figure 8.32
A lateral view of the right foot.
 1. Tendon of tibialis anterior m.
 2. Tendon of extensor hallucis longus m.
 3. Tendon of peroneus tertius m.
 4. Tendon of extensor digitorum longus m.
 5. Superior extensor retinaculum
 6. Inferior extensor retinaculum
 7. Lateral malleolus of fibula
 8. Extensor digitorum brevis m.
 9. Tendon of peroneus longus m.
10. Tendon of peroneus brevis m.
11. Calcaneus
12. Fifth metatarsal bone

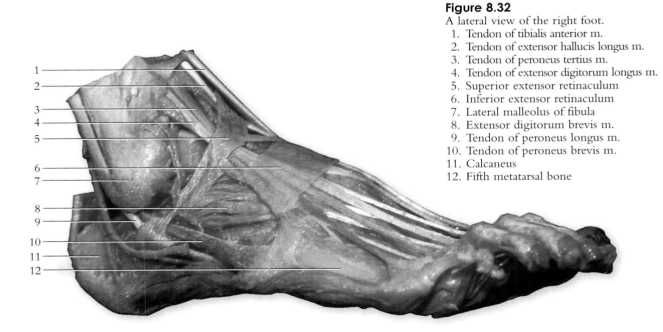

9 Nervous System

The nervous system is anatomically divided into the **central nervous system (CNS)**, which includes the **brain** and **spinal cord**, and the **peripheral nervous system (PNS)**, which includes the **cranial nerves**, arising from the brain, and the **spinal nerves**, arising from the spinal cord (fig. 9.1). The **autonomic nervous system (ANS)** is a functionally distinct division of the nervous system devoted to regulation of involuntary activities in the body. The ANS is made up of specific portions of the CNS and PNS.

The brain and spinal cord are the centers for integration and coordination of information. Conveyed as **nerve impulses**, information to and from the brain and spinal cord travels through **nerves**. Nerves are similar to electrical conducting wires. Nerve impulses are sent from the brain in the form of electrical signals along **motor nerves** to the receiving organs, which then translate the signal into some specific function. For example, the motor impulses conducted from the brain to the muscles of the forearm that serve the hand cause the fingers to move as the muscles are contracted. **Sensory nerves** conduct action potentials (nerve impulses) in the opposite direction—from the receptor site to the CNS. For example, a pinprick on the skin produces a sensory impulse along a sensory nerve that the brain interprets as a painful sensation.

Neurons and **neuroglia** are the two cell types that make up nervous tissue. Neurons are specialized to respond to physical and chemical stimuli, conduct impulses, and release specific chemical regulators, called **neurotransmitters**. Although neurons vary considerably in size and shape, they have three principal components: a **cell body, dendrites**, and an **axon** (fig. 9.3). In a typical neuron connection, the axon of one neuron **synapses** (joins) on the cell body or dendrites of a neighboring neuron. Axons vary in length from a few millimeters in the CNS to over a meter in the PNS. Long axons are generally **myelinated** with **neurolemmocytes (Schwann cells)** in the PNS, and many of the short axons are myelinated with **oligodendrocytes** in the CNS. **Neurofibril nodes (nodes of Ranvier)** are segments in the **myelin sheath**. The end of the axon at the synapse is called the **axon terminal**.

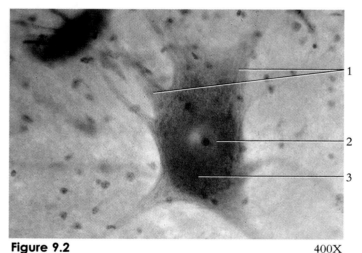

Figure 9.2   400X
A photomicrograph of a neuron.
1. Cytoplasmic extensions   3. Cell body of neuron
2. Nucleus

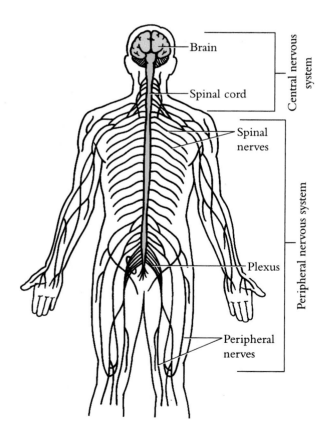

Figure 9.1
Divisions of the nervous system.

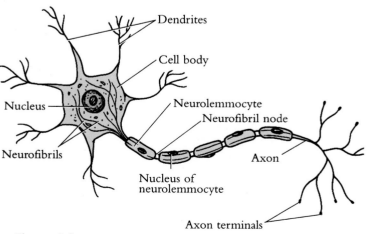

Figure 9.3
Structure of a myelinated neuron.

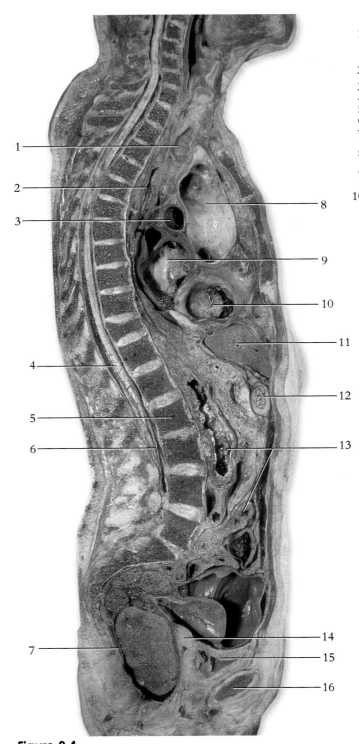

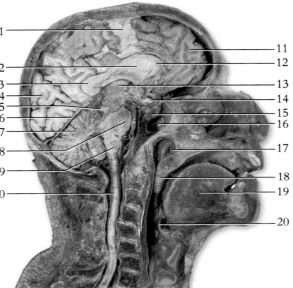

Figure 9.5

A sagittal section of the head and neck.

1. Remnent of falx cerebri
2. Septum pellucindum
3. Genu of corpus callosum
4. Occipital lobe of cerebrum
5. Corpora quadrigemina
6. Falx cerebelli
7. Cerebellum
8. Pons
9. Medulla oblongata
10. Spinal cord
11. Frontal lobe of cerebrum
12. Splenium of corpus callosum
13. Thalamus
14. Optic chiasma
15. Pituitary gland
16. Sphenoidal sinus
17. Pharyngeal opening of auditory tube
18. Uvula
19. Tongue
20. Epiglottis

Figure 9.4

A sagittal section of the female trunk.

1. Trachea
2. Esophagus
3. Pulmonary artery
4. Spinal cord
5. Body of lumbar vertebra
6. Cauda equina
7. Rectum
8. Aorta
9. Right atrium
10. Right ventricle
11. Liver
12. Stomach
13. Small Intestine
14. Uterus
15. Urinary bladder
16. Pubic bone

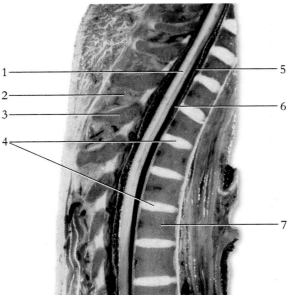

Figure 9.6

A sagittal section of the spinal column in the cervical and superior thoracic regions.

1. Spinal cord
2. Semispinalis cervicis muscle
3. Spinous process
4. Intervertebral discs
5. Dura mater
6. Subarachnoid space
7. Body of vertebra

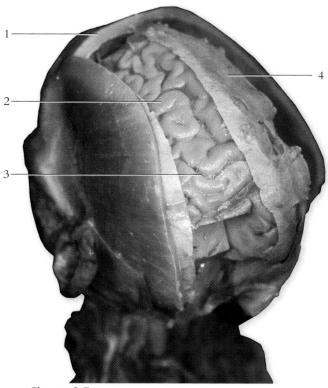

Figure 9.7
A sectioned cranium exposing the meninges and cerebrum.
1. Skull
2. Cerebral gyrus
3. Arachnoid
4. Dura mater

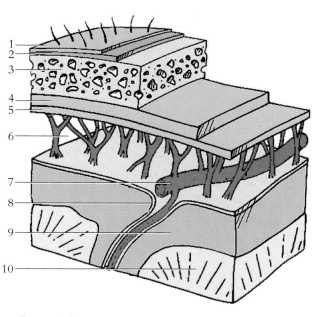

Figure 9.8
The relationship of the meninges to the skull and the cerebrum.
1. Skin of scalp
2. Galea aponeurotica
3. Bone of cranium
4. Dura mater
5. Arachnoid
6. Subarachnoid space
7. Blood vessel
8. Pia mater
9. Cerebral cortex
   (gray matter)
10. Cerebral medulla
    (white matter)

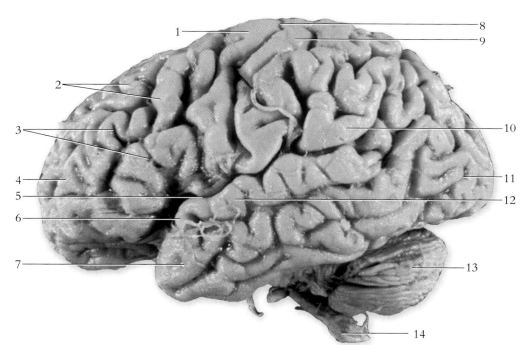

Figure 9.9
A lateral view of the brain.
1. Primary motor
   cerebral cortex
2. Gyri
3. Sulci
4. Frontal lobe of cerebrum
5. Lateral sulcus
6. Olfactory cerebral cortex
7. Temporal lobe of cerebrum
8. Central sulcus
9. Primary motor
   cerebral cortex
10. Parietal lobe of cerebrum
11. Occipital lobe of cerebrum
12. Auditory cerebral cortex
13. Cerebellum
14. Medulla oblongata

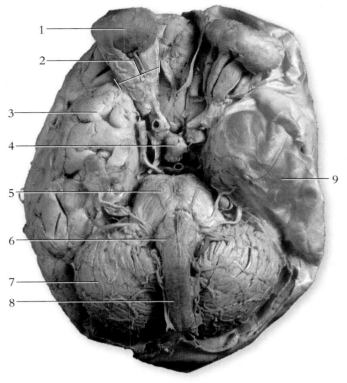

Figure 9.10

An inferior view of the brain with the eyes and
part of the meninges still intact.

1. Eyeball
2. Muscles of the eye
3. Temporal lobe of cerebrum
4. Pituitary gland
5. Pons
6. Medulla oblongata
7. Cerebellum
8. Spinal cord
9. Dura mater

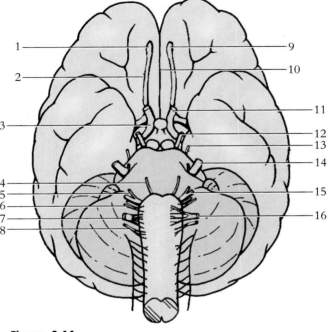

Figure 9.11

A diagram of the inferior of the brain showing the cranial nerves.

1. Olfactory bulb
2. Olfactory tract
3. Optic tract
4. Abducens nerve
5. Facial nerve
6. Glossopharyngeal nerve
7. Vagus nerve
8. Accessory nerve
9. Olfactory nerve
10. Longitudinal cerebral
    fissure
11. Optic nerve
12. Oculomotor nerve
13. Trochlear nerve
14. Trigeminal nerve
15. Vestibulocochlear nerve
16. Hypoglossal nerve

Figure 9.12

Cranial nerves and blood
supply to the brain.

1. Internal carotid artery
2. Cerebral arterial circle
   (circle of Willis)
3. Trigeminal nerve
4. Abducens nerves
5. Vestibulocochlear nerve
6. Olfactory tract
7. Optic nerve
8. Optic chiasma
9. Oculomotor nerve
10. Trochlear nerve
11. Trigeminal nerve
12. Facial nerve
13. Glossopharyngeal nerve
14. Vagus nerve
15. Vertebral artery

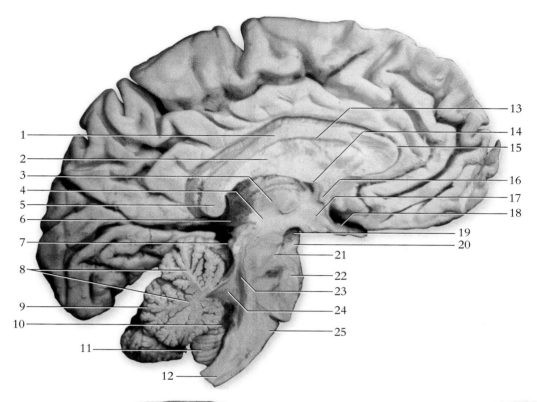

Figure 9.13
A sagittal view of the brain.
1. Body of corpus callosum
2. Crus of fornix
3. Third ventricle
4. Posterior commissure
5. Splenium of corpus callosum
6. Pineal body
7. Inferior colliculus
8. Arbor vitae of cerebellum
9. Vermis of cerebellum
10. Choroid plexus of fourth ventricle
11. Tonsilla of cerebellum
12. Medulla oblongata
13. Septum pellucidum (cut)
14. Intraventricular foramen
15. Genu of corpus callosum
16. Anterior commissure
17. Hypothalmus
18. Optic chiasma
19. Oculomotor nerve
20. Cerebral peduncle
21. Midbrain
22. Pons
23. Mesencephalic (cerebral) aqueduct
24. Fourth ventricle
25. Pyramid

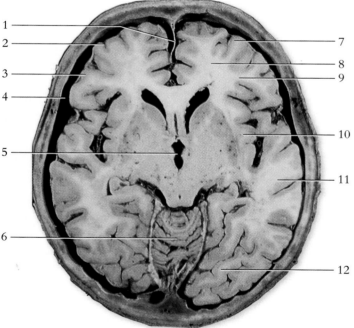

Figure 9.14
Transaxial section of the skull and brain.
1. Falx cerebri (septum of dura matter)
2. Sulcus
3. Gyrus
4. Subdural space
5. Mesencephalic aqueduct (cerebral aqueduct)
6. Cerebellum
7. Cerebral cortex (gray matter)
8. Cerebral medulla (white matter)
9. Frontal lobe
10. Insula
11. Temporal lobe
12. Occipital lobe

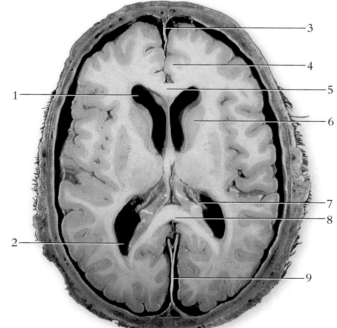

Figure 9.15
Transaxial section of the skull and brain.
1. First ventricle
2. Second ventricle
3. Falx cerebri (septum of dura mater)
4. Cingulate gyrus
5. Genu of corpus callosum
6. Caudate nucleus
7. Choroid plexus
8. Splenium of corpus callosum
9. Falx cerebri (septum of dura mater)

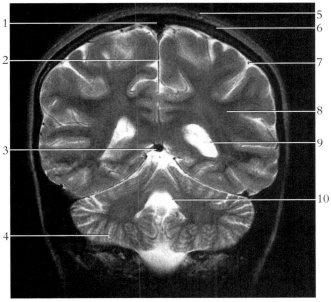

Figure 9.16
Coronal MRI brain scan.
1. Cerebrospinal fluid
2. Longitudinal cerebral fissure
3. Third ventricle
4. Cerebellum
5. Skull
6. Dura mater
7. Cerebral cortex
8. Cerebral medulla
9. Lateral ventricle
10. Fourth ventricle

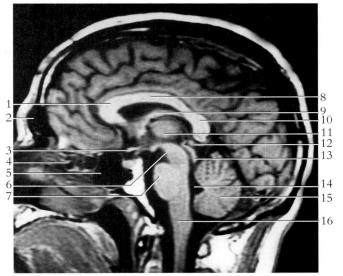

Figure 9.17
MRI sagittal section through the skull.
1. Genu of corpus callosum
2. Frontal sinus
3. Pituitary gland
4. Ethmoidal sinus
5. Sphenoidal sinus
6. Tegmentum (midbrain)
7. Pons
8. Body of corpus callosum
9. Fornix
10. Splenium of corpus callosum
11. Thalamus
12. Pineal gland
13. Superior and inferior colliculi
14. Fourth ventricle
15. Cerebellum
16. Medulla oblongata

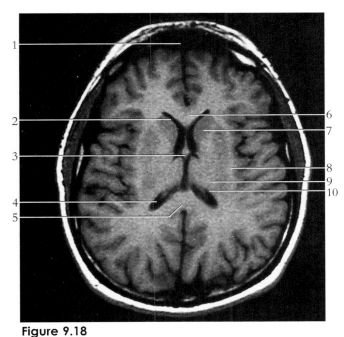

Figure 9.18
MRI transaxial section through the brain.
1. Frontal sinus
2. Frontal horn of lateral ventricle
3. Body of fornix
4. Posterior horn of lateral ventricle
5. Splenium of corpus callosum
6. Genu of corpus callosum
7. Head of caudate nucleus
8. External capsule
9. Thalamus
10. Choroid plexus

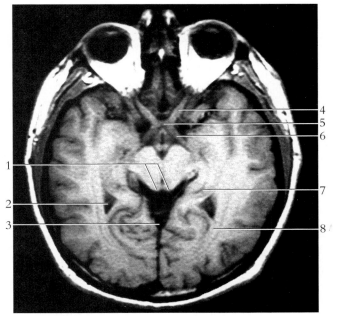

Figure 9.19
MRI transaxial section showing visual pathways.
1. Superior colliculi
2. Lateral ventricle
3. Third ventricle
4. Optic nerve
5. Optic chiasma
6. Optic tract
7. Lateral geniculate body
8. Calcarine tracts (optic radiation)

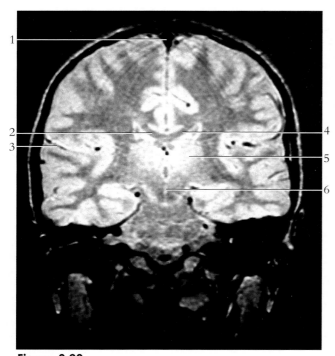

Figure 9.20
A MRI coronal section through the thalamus.

1. Superior sagittal sinus
2. Lateral ventricle
3. Lateral fissure
4. Corpus callosum
5. Thalamus
6. Third ventricle

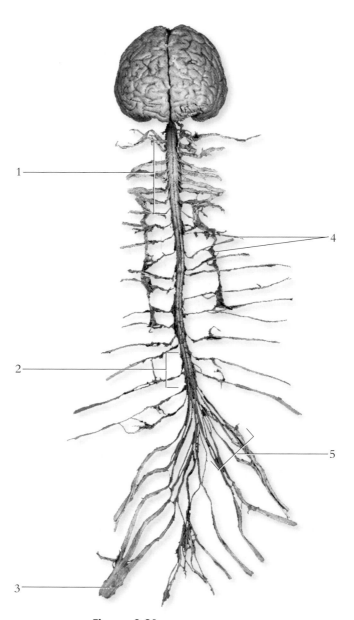

Figure 9.21
Anterior surface of the brain and spinal cord with meninges removed.

1. Cervical enlargement
2. Lumbar enlargement
3. Sciatic nerve
4. Sympathetic ganglia
5. Cauda equina

Figure 9.22
A posterior view of the lower spinal cord.

1. Spinal cord
2. Dura mater (cut)
3. Posterior (dorsal) root of spinal nerve
4. Cauda equina
5. Filum terminale

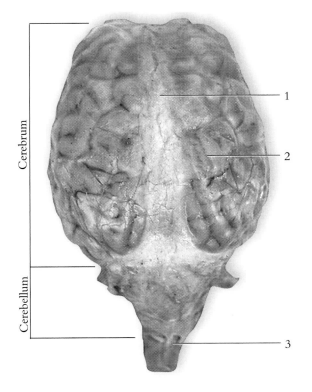

1 Cerebrum

2

3

Cerebellum

Figure 9.23
Sheep brain, dorsal view.
  1. Dura mater covering cerebral longitudinal fissure
  2. Arachnoid
  3. Medulla oblongata

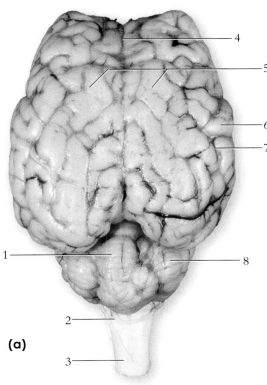

4

5

6

7

1

8

2

3

(a)

Frontal lobe

Longitudinal
cerebral fissure

Sulci

Gyri

Parietal
lobe

Occipital
lobe

Vermis of
cerebellum

Medulla oblongata

Cerebrum

Cerebellum

(b)

Figure 9.24
Sheep brain, dorsal view. (a) photograph; (b) diagram.
  1. Vermis
  2. Medulla oblongata
  3. Spinal Cord
  4. Longitudinal cerebral fissure
  5. Cerebral hemipheres
  6. Gyrus
  7. Sulcus
  8. Cerebellar hemisphere

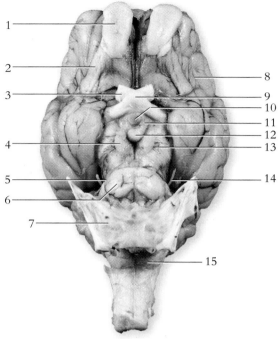

Figure 9.25

Ventral view of sheep brain with dura mater cut and reflected.
1. Olfactory bulb
2. Olfactory tract
3. Optic nerve
4. Trigeminal nerve
5. Oculomotor nerve
6. Pons
7. Dura mater (cut)
8. Pia mater (adhering to brain)
9. Optic chiasma
10. Position of pituitary stock
11. Tuber cinereum
12. Mammillary body
13. Cerebral penduncle
14. Trochlear nerve
15. Medulla oblongata

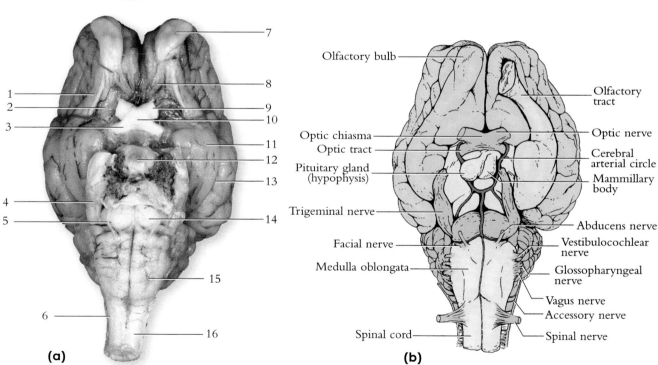

(a) (b)

Figure 9.26

Sheep brain, venral view. (a) photograph; (b) diagram.

1. Lateral olfactory band
2. Olfactory trigone
3. Optic tract
4. Trigeminal nerve
5. Abducens nerve
6. Accessory nerve
7. Olfactory bulb
8. Medial olfactory band
9. Optic nerve
10. Optic chiasma
11. Pyriform lobe
12. Pituitary gland (hypophysis)
13. Rhinal sulcus
14. Pons
15. Medulla oblongata
16. Spinal cord

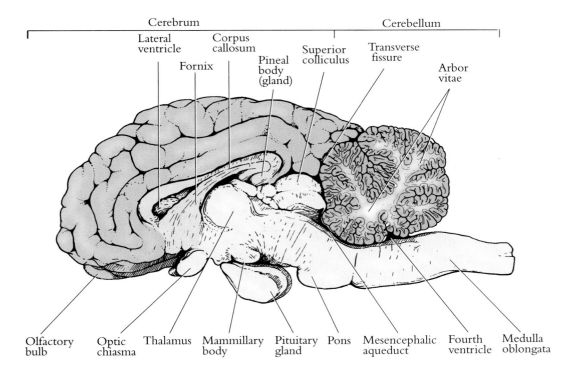

Figure 9.27
Sheep brain, sagittal view.

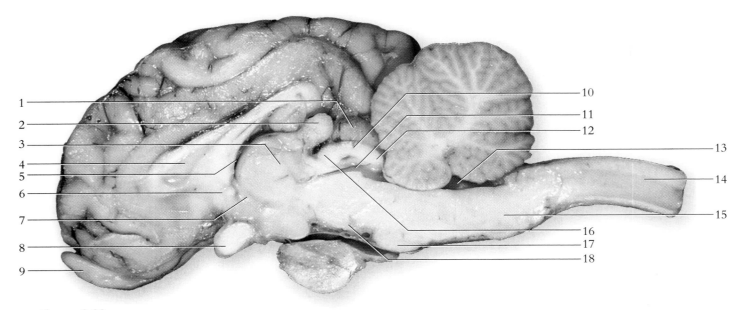

Figure 9.28
Cat brain, right sagittal view.
1. Superior colliculus
2. Pineal body (gland)
3. Intermediate mass
4. Septum pellucidum
5. Interventicular foramen
   (foramen of Monro)
6. Anterior commissure
7. Third ventricle
8. Optic chiasma
9. Olfactory bulb
10. Corpora quadrigemina
11. Mesencephalic (cerebral) aqueduct
12. Inferior colliculus
13. Fourth ventricle
14. Spinal cord
15. Medulla oblongata
16. Posterior commissure
17. Pons
18. Cerebral peduncle

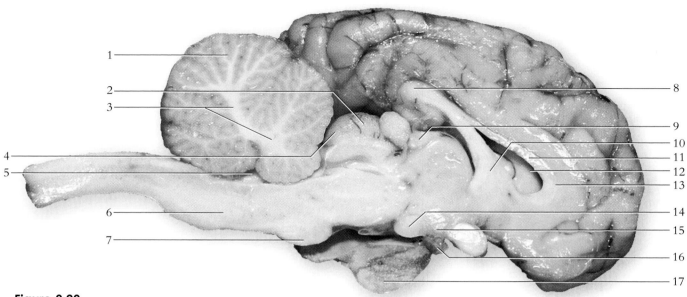

Figure 9.29
Cat brain, left sagittal view.
1. Cerebellum
2. Superior colliculus
3. Arbor vitae
4. Inferior colliculus
5. Fourth ventricle
6. Medulla oblongata

7. Pons
8. Splenium of corpus callosum
9. Habenular trigone
10. Fornix
11. Body of corpus callosum
12. Lateral ventricle

13. Genu of corpus callosum
14. Mammillary body
15. Tuber cinereum
16. Pituitary stalk
17. Pituitary gland (hypophysis)

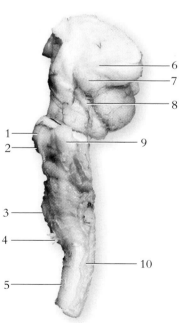

Figure 9.30
Brainstem, lateral view.
1. Pons
2. Abducens nerve
3. Medulla oblongata
4. Hypoglossal nerve
5. Spinal cord
6. Lateral geniculate body
7. Medial geniculate body
8. Trochlear nerve
9. Trigeminal nerve
10. Accessory nerve

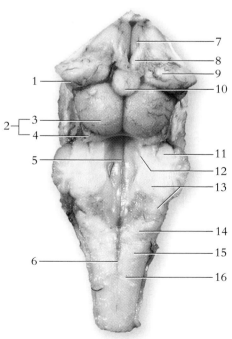

Figure 9.31
Brainstem, dorsal view.
1. Medial geniculate body
2. Corpora quadrigemina
3. Superior colliculus
4. Inferior colliculus
5. Fourth ventricle
6. Dorsal median sulcus
7. Intermediate mass
8. Habenular trigone
9. Thalmus
10. Pineal gland
11. Middle cerebellar peduncle
12. Anterior cerebellar peduncle
13. Posterior cerebellar peduncle
14. Tuberculum cuneatum
15. Fasciculus gracilis
16. Fasciculus cuneatus

The endocrine system works closely with the nervous system to regulate and integrate body processes and maintain homeostasis. The nervous system regulates body activities through the action of electrochemical impulses that are transmitted by means of neurons, resulting in rapid, but usually brief responses. By contrast, the endocrine system is composed of glands (fig. 10.1) scattered throughout the body that release chemical substances called **hormones** into the bloodstream. These hormones dissipate in the blood and travel throughout the entire body to act on **target tissues**, where they have a slow but relatively long-lasting effect. Neurological responses are measured in milliseconds, but hormonal action requires seconds or days to elicit a response. Some hormones may have an effect that lasts for minutes and others for weeks or months.

The endocrine system and nervous system are closely coordinated in autonomically controlling the functions of the body. The **pituitary gland**, located in the brain, regulates the activity of most other endocrine glands. Located immediately between the pituitary and the rest of the brain is the **hypothalamus**. The hypothalamus serves as an intermediate between the nervous centers of the brain and the pituitary gland, correlating the activity of the two systems. Furthermore, certain hormones may stimulate or inhibit the activities of the nervous system.

Other organs of the endocrine system include the **thyroid gland** and **parathyroid glands**, located in the neck. The **adrenal glands** and **pancreas** are located in the abdominal region. The **ovaries** of the female are located in the pelvic cavity, whereas the **testes** of the male are located in the scrotum. Even the **placenta** serves as an endocrine organ for the developing fetus and has some hormonal influence upon the pregnant woman.

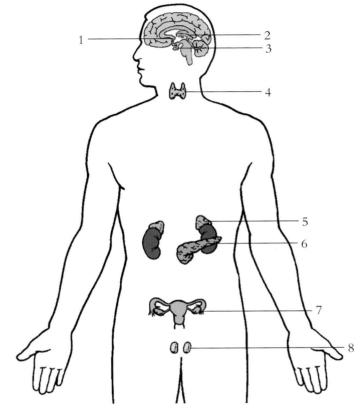

Figure 10.1
Principal endocrine glands.
1. Hypothalamus
2. Pineal body
3. Pituitary gland
4. Thyroid and parathyroid glands
5. Adrenal (suprarenal) gland
6. Pancreas
7. Ovary
8. Testis

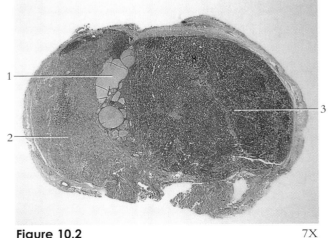

Figure 10.2   7X
Pituitary gland.
1. Pars intermedia (adenohypophysis)
2. Pars nervosa (neurohypophysis)
3. Pars distalis (adenohypophysis)

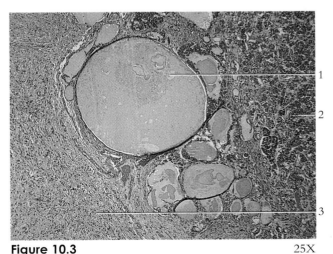

Figure 10.3   25X
Pituitary gland.
1. Pars intermedia
2. Pars distalis
3. Pars nervosa

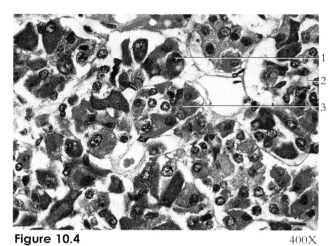

Figure 10.4   400X
Pars distalis of the pituitary gland.
   1. Basophil
   2. Chromophobe
   3. Acidophil

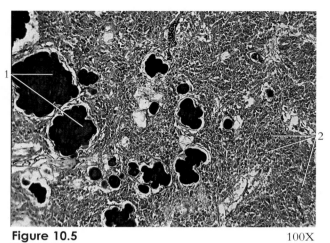

Figure 10.5   100X
Pineal gland.
   1. Brain sand
   2. Pinealocytes

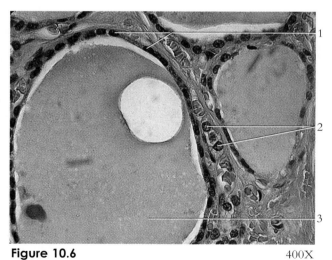

Figure 10.6   400X
Thyroid gland.
   1. Follicle cells
   2. C cells
   3. Colloid within follicle

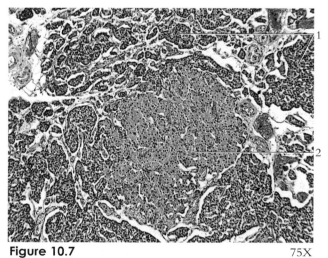

Figure 10.7   75X
Parathyroid gland.
   1. Chief cells
   2. Cluster of oxyphil cells

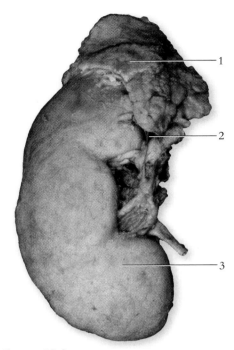

Figure 10.8
The adrenal (suprarenal) gland.
1. Adrenal gland
2. Inferior suprarenal artery
3. Kidney

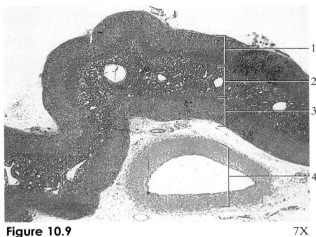

Figure 10.9   7X
Adrenal gland.
1. Adrenal cortex
2. Adrenal medulla
3. Adrenal cortex
4. Blood vessel

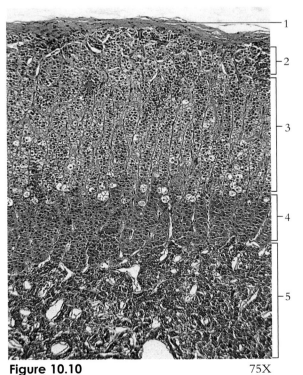

Figure 10.10   75X
Adrenal gland.
1. Capsule
2. Zona glomerulosa
   (adrenal cortex)
3. Zona fasiculata
   (adrenal cortex)
4. Zona reticularis
   (adrenal cortex)
5. Adrenal medulla

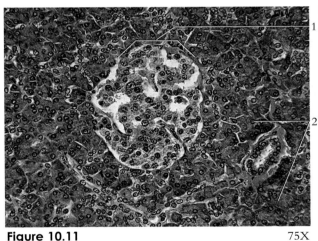

Figure 10.11   75X
Pancreatic islet (islet of Langerhans).
1. Pancreatic islet (endocrine pancreas)
2. Acini (exocrine pancreas)

The nervous and endocrine systems convey information from the brain to all parts of the body to enable a person to interact with both the external and internal environments and to maintain homeostasis. The sense organs, in contrast, convey information from the outside world (and inside world of the body) back to the brain. This includes a wide range of information such as temperature, brightness, sound, flavor, and balance.

The sense organs are actually extensions of the nervous system that allow us to autonomically respond or conscientiously perceive our internal and external environments. A stimulus excites a sense organ which then transduces the stimulus to an electrical (nerve) impulse. Sensory nerves transmit the impulse (sensation) to the brain to be perceived and acted upon. Ultimately, it is the brain which actually feels, sees, hears, tastes, and smells.

The **eyes** are the organs of visual sense. The eyes refract (bend) and focus the incoming light waves onto the sensitive **photoreceptors** (**rods** and **cones**) at the back of each eye. Nerve impulses from the stimulated photoreceptors are conveyed along visual pathways to the occipital lobes of the cerebrum, where visual sensations are perceived.

The eyeball consists of the fibrous tunic, which is divided into the **sclera** and **cornea**; the vascular tunic, which consists of the choroid, the **ciliary body**, and the **iris**; and the internal tunic, or **retina**, which consists of an outer pigmented layer and an inner nervous layer. The eye contains an anterior cavity between the lens and the retina. The anterior cavity is subdivided into an anterior chamber in front of the iris and a posterior chamber behind the iris. **Aqueous humor** fills both of these chambers. The posterior cavity (also called the vitreous chamber) contains vitreous humor.

The **ear** is the organ of hearing and equilibrium (balance). It contains receptors that respond to movements of the head and receptors that convert sound waves into nerve impulses. Impulses from both receptor types are transmitted through the vestibulocochlear (VIII) cranial nerve to the brain for interpretation.

The ear consists of the three principal regions: the **outer ear**, the **middle ear**, and the **inner ear**. The outer ear consists of the **auricle** and the **external auditory canal**. The middle ear contains the auditory ossicles (**malleus**, **incus**, and **stapes**). The inner ear contains the **spiral organ** (organ of Corti) in the **cochlea** for hearing, and the **semicircular canals** and the vestibular organs for equilibrium.

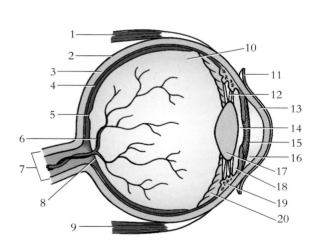

Figure 11.1
Structure of the eye.
1. Superior rectus m.
2. Sclera
3. Choroid
4. Retina
5. Fovea centralis
6. Central vessels
7. Optic nerve
8. Optic disc
9. Inferior rectus m.
10. Posterior cavity (contains vitreous humor)
11. Conjuctiva
12. Suspensory ligament
13. Cornea
14. Pupil
15. Iris
16. Anterior chamber
17. Lens
18. Posterior chamber
19. Ciliary body
20. Ora serrata

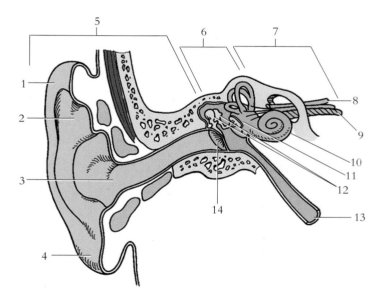

Figure 11.2
Structure of the ear.
1. Helix
2. Auricle
3. External auditory canal
4. Earlobe
5. Outer ear
6. Middle ear
7. Inner ear
8. Facial nerve
9. Vestibulocochlear nerve
10. Cochlea
11. Vestibular (oval) window
12. Auditory ossicles
13. Auditory tube
14. Tympanic membrane

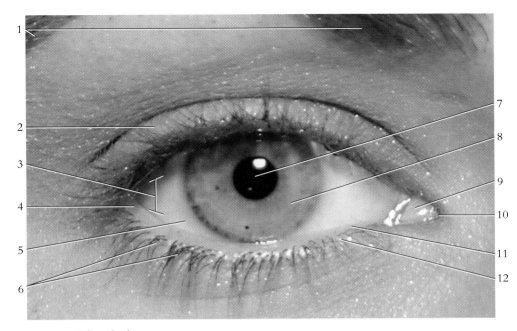

Figure 11.3
The surface anatomy of the eye.
1. Eyebrow
2. Superior eyelid (palpebra)
3. Palpebra commisure
4. Lateral canthus
5. Sclera
6. Eyelashes
7. Pupil
8. Iris
9. Lacrimal caruncle
10. Medial commisure
11. Conjuctiva
12. Inferior eyelid (palpebra)

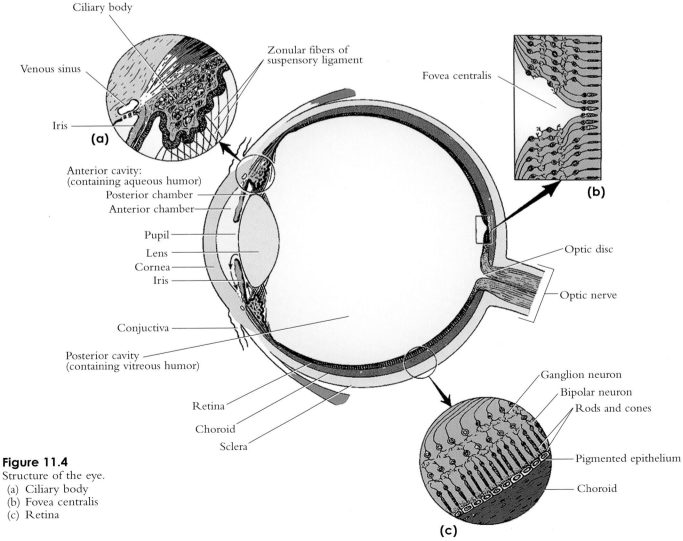

Figure 11.4
Structure of the eye.
(a) Ciliary body
(b) Fovea centralis
(c) Retina

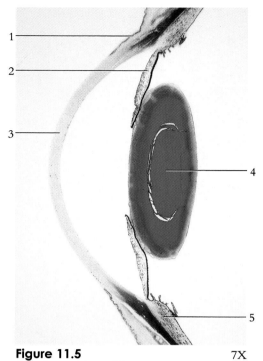

Figure 11.5   7X
Anterior portion of the eye.
1. Conjunctivia   4. Lens
2. Iris   5. Ciliary body
3. Cornea

Figure 11.6   250X
Retina.
1. Retina   3. Choroid
2. Rods and cones   4. Sclera

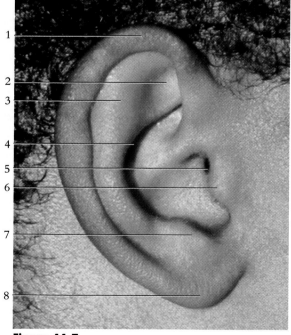
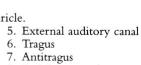

Figure 11.7
The surface anatomy of the auricle.
1. Helix   5. External auditory canal
2. Triangular fossa   6. Tragus
3. Antihelix   7. Antitragus
4. Concha   8. Earlobe

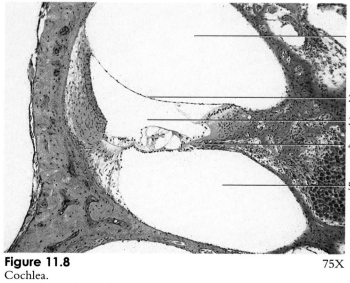

Figure 11.8   75X
Cochlea.
1. Scala vestibuli   4. Basilar membrane
2. Vestibular membrane   5. Scala tympani
3. Cochlear duct

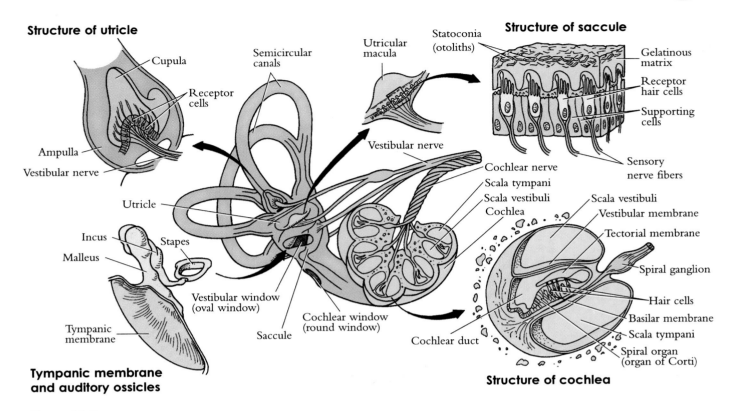

Structure of utricle

Cupula

Receptor cells

Ampulla

Vestibular nerve

Semicircular canals

Utricular macula

Utricle

Incus

Malleus

Stapes

Vestibular nerve

Vestibular window (oval window)

Cochlear window (round window)

Saccule

Tympanic membrane

Tympanic membrane and auditory ossicles

Statoconia (otoliths)

Structure of saccule

Gelatinous matrix

Receptor hair cells

Supporting cells

Sensory nerve fibers

Cochlear nerve

Scala tympani

Scala vestibuli

Cochlea

Scala vestibuli

Vestibular membrane

Tectorial membrane

Spiral ganglion

Hair cells

Basilar membrane

Scala tympani

Spiral organ (organ of Corti)

Cochlear duct

Structure of cochlea

Figure 11.9
Structures of the middle ear and inner ear. The tympanic membrane and auditory ossicles (malleus, incus, stapes) are structures of the middle ear. The vestibular organs (utricle, saccule, semicircular canals) and cochlea (containing the spiral organ) are structures of the inner ear.

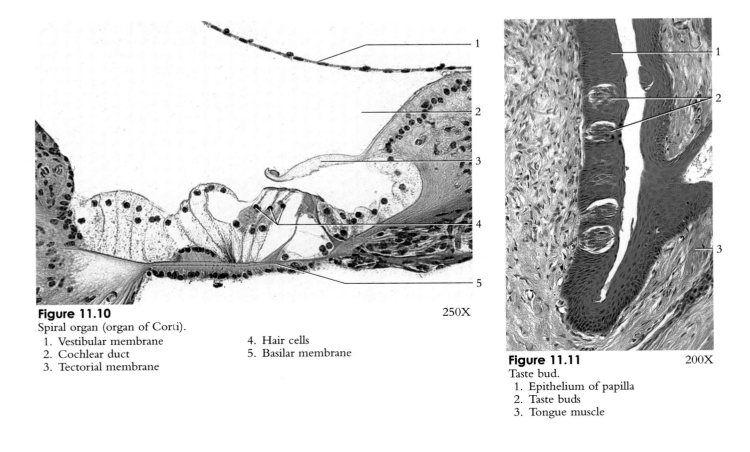

Figure 11.10
Spiral organ (organ of Corti).   250X
1. Vestibular membrane
2. Cochlear duct
3. Tectorial membrane
4. Hair cells
5. Basilar membrane

Figure 11.11   200X
Taste bud.
1. Epithelium of papilla
2. Taste buds
3. Tongue muscle

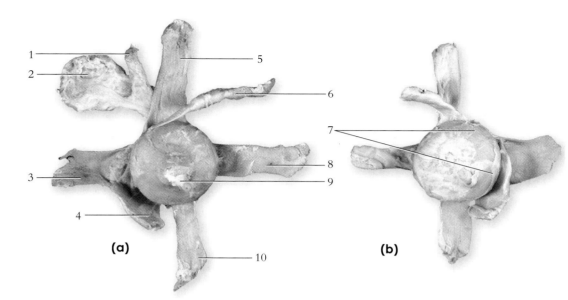

(a)

(b)

Figure 11.12
Extrinsic eye muscles of a cat.
(a) A posterior view with the optic nerve intact.
(b) A posterior view with the optic nerve removed.
1. Levator palpebrae superioris m.
2. Lacrimal gland
3. Lateral rectus m.
4. Inferior oblique m.
5. Superior rectus m.
6. Superior oblique m.
7. Retractor bulbi
8. Medial rectus m.
9. Optic nerve
10. Inferior rectus m.

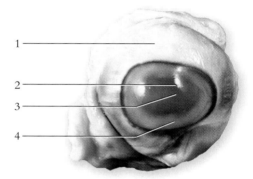

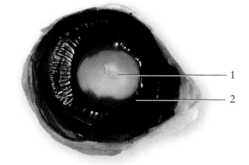

Figure 11.13
Superficial view of the anterior eyeball of a cat.
1. Sclera
2. Cornea
3. Pupil (dark opening)
4. Iris

Figure 11.14
Anterior view of the eyeball with the lens in natural position.
1. Lens
2. Iris

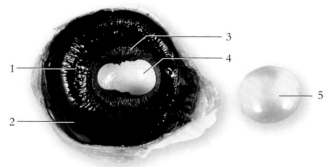

Figure 11.15
Tunics of the cat eyeball.
1. Sclera
2. Coroid
3. Retina
4. Tapetum lucidum
5. Optic disc

Figure 11.16
Internal anatomy of cat eye.
1. Ciliary body
2. Ora serrata
3. Iris
4. Pupil
5. Lens (removed)

The circulatory system consists of the **blood, heart**, and **vessels**, each of which is essential to the life of a complex multicellular organism. Blood, a specialized connective tissue, consists of **formed elements** (erythrocytes, leukocytes, and thrombocytes) that are suspended and carried in the **plasma**. These formed elements function in transport, immunity, and blood-clotting mechanisms.

The heart is enclosed in a **pericardial sac** within the thoracic cavity. The wall of the heart consists of the **epicardium, myocardium**, and **endocardium**. The **right atrium** of the heart receives blood from the superior vena cava and inferior vena cava, and the **right ventricle** pumps blood into the **pulmonary trunk** to the **pulmonary arteries**. The **left atrium** receives blood from the **pulmonary veins** and pumps blood into the **left ventricle**. The left ventricle pumps blood into the **aorta**.

There are four heart valves that prohibit the backflow of blood: 1) The **right atrioventricular valve (right tricuspid valve)** is located between the right atrium and the right ventricle; 2) the **pulmonary valve (pulmonary semilunar valve)** is located between the right ventricle and the pulmonary trunk; 3) the **left atrioventricular valve (left bicuspid, or mitral valve)** is located between the left atrium and the left ventricle; and, 4) the **aortic valve (aortic semilunar valve)** is located between the left ventricle and the ascending aorta.

The **systemic arteries** arise from the aorta or branches of the aorta and transport blood away from the heart to smaller vessels called **arterioles**. From arterioles, the blood enters **capillaries** where diffusion with the surrounding cells may occur. Capillaries converge forming **venules**, which in turn converge forming larger vessels called **veins**. Veins are vessels that transport blood toward the heart.

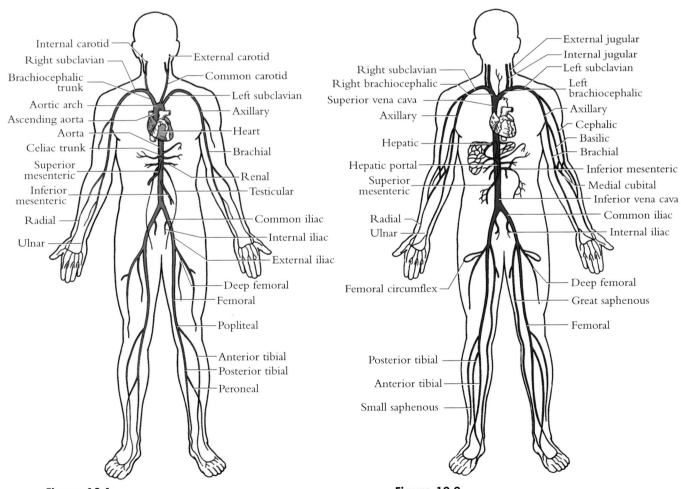

Figure 12.1
Principal arteries of the body.

Figure 12.2
Principal veins of the body.

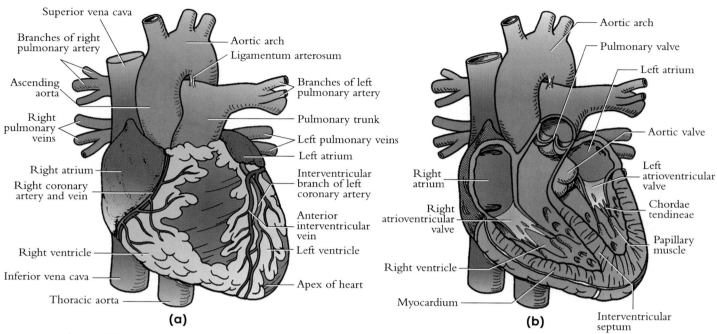

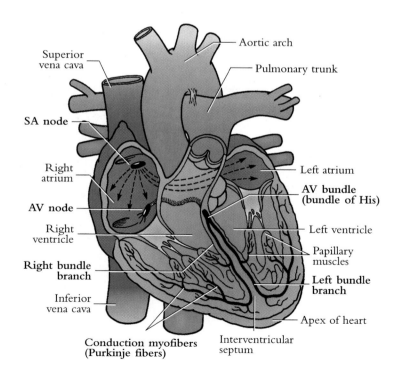

Figure 12.3

The structure of the heart.
(a) An anterior view
(b) an internal view.

Figure 12.4

Conduction system of the heart, indicated in bold type.

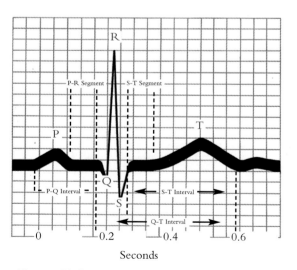

Figure 12.5

Normal electrocardiogram (ECG).

Table 12.1 Valves of the heart.

Valve	Location	Structure and Function
Right atrioventricular valve (tricuspid valve)	Between right atrium and right ventricle	Consists of three cusps that prevent backflow of blood during systole (ventricular contraction)
Pulmonary valve (pulmonary semilunar valve)	Entrance to pulmonary trunk	Consists of three partial-moon-shaped flaps that prevent backflow of blood during diastole (ventricular relaxation)
Left atrioventricular valve (bicuspid valve or mitral valve)	Between left atrium and left ventricle	Consists of two cusps that prevent backflow of blood during systole
Aortic valve (aortic semilunar valve)	Entrance to ascending aorta	Consists of three partial-moon-shaped flaps that prevent backflow of blood during diastole

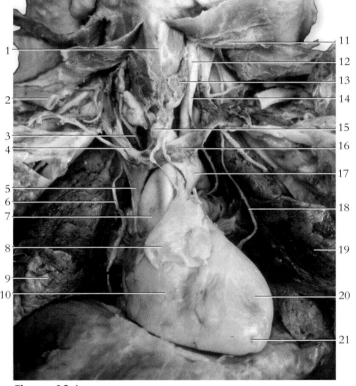

Figure 12.6
An anterior view of the heart and associated structures
1. Thyroid cartilage of larynx
2. First rib (cut)
3. Right vagus nerve
4. Right brachiocephalic vein
5. Superior vena cava
6. Right phrenic nerve
7. Ascending aorta
8. Pericardium (cut)
9. Right lung
10. Right ventricle of heart
11. Sternohyoid (cut and reflected)
12. Left common carotid artery
13. Thyroid gland (cut)
14. Left vagus nerve
15. Brachiocephalic artery
16. Left brachiocephalic artery
17. Aortic arch
18. Left phrenic nerve
19. Left lung
20. Left ventricle of heart
21. Apex of heart

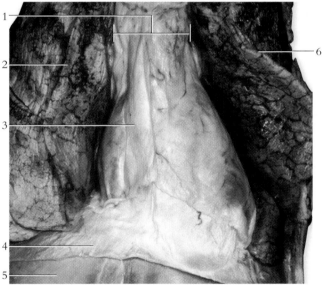

Figure 12.7
The position of the heart within the pericardium.
1. Mediastinum
2. Right lung
3. Pericardium
4. Diaphragm
5. Liver
6. Left lung

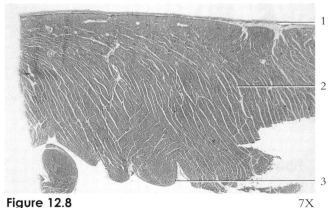

Figure 12.8   7X
Wall of the heart.
1. Epicardium
2. Myocardium
3. Endocardium

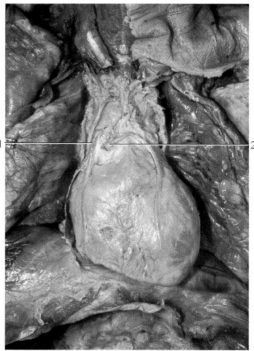

Figure 12.9

A double coronary artery bypass surgery. Several
vessels may be used in the autotransplant,
including the internal thoracic artery and
the great saphenous vein.
1. A graft to the ascending aorta
2. A graft to the left coronary artery

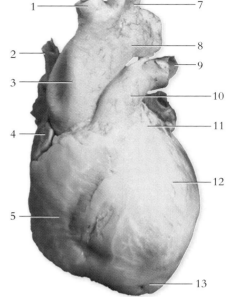

Figure 12.10

An anterior view of the heart
and great vessels.
1. Brachiocephalic trunk
2. Superior vena cava
3. Ascending aorta
4. Right atrium
5. Right ventricle
6. Left common carotid artery
7. Left subclavian artery
8. Aortic arch
9. Pulmonary artery
10. Pulmonary trunk
11. Left atrium
12. Left ventricle
13. Apex of heart

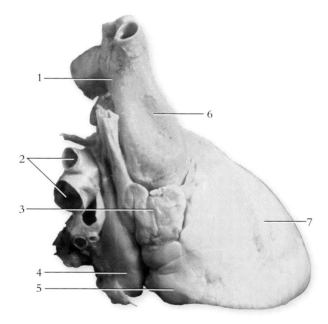

Figure 12.11

A posterior view of the heart.
1. Aortic arch   5. Right ventricle
2. Pulmonary arteries   6. Ascending aorta
3. Right atrium   7. Left ventricle
4. Inferior vena cava

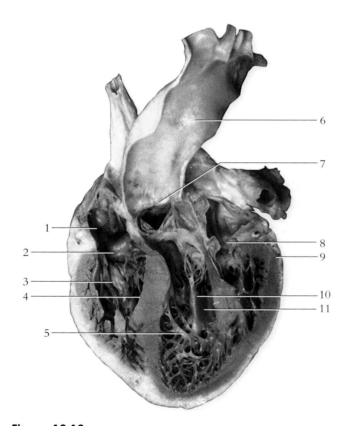

Figure 12.12

The internal structure of the heart.
1. Right atrium   7. Aortic valve
2. Right atrioventricular valve   8. Left atrioventricular valve
3. Right ventricle   9. Myocardium
4. Interventricular septum   10. Papillary muscle
5. Trabeculae carneae   11. Left ventricle
6. Ascending aorta

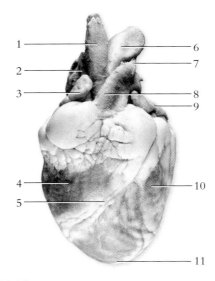

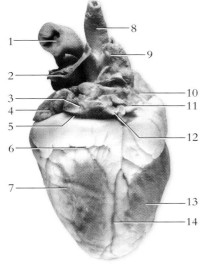

Figure 12.13
Ventral view of mammalian (sheep) heart .
1. Brachiocephalic artery
2. Cranial vena cava
3. Right auricle of right atrium
4. Right ventricle
5. Interventicular groove
6. Ascending aorta
7. Ligamentum arteriosum
8. Pulmonary trunk
9. Left auricle of left atrium
10. Left ventricle
11. Apex of heart

Figure 12.14
Dorsal view of mammalian (sheep) heart .
1. Aorta
2. Pulmonary artery
3. Pulmonary vein
4. Left auricle
5. Left atrium
6. Atrioventricular groove
7. Left ventricle
8. Brachiocephalic artery
9. Cranial vena cava
10. Right auricle
11. Right atrium
12. Pulmonary vein
13. Right ventricle
14. Interventricular groove

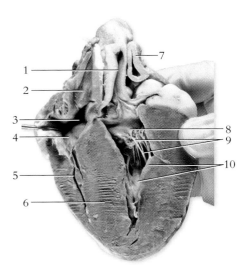

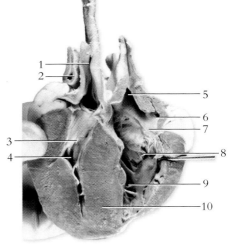

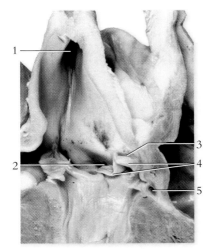

Figure 12.15
Coronal section of the
mammalian (sheep) heart .
1. Aorta
2. Cranial vena cava
3. Right atrium
4. Right atrioventricular (tricuspid) valve
5. Right ventricle
6. Interventricalar septum
7. Pulmonary artery
8. Left atrioventricular (bicuspid) valve
9. Chordae tendineae
10. Papillary muscles

Figure 12.16
Coronal section of the mammalian
(sheep) heart showing the valves.
1. Opening of the brachiocephalic artery
2. Pulmonary artery
3. Left atrioventricular (bicuspid) valve
4. Left ventricle
5. Opening of cranial vena cava
6. Opening of coronary sinus
7. Right atrium
8. Right atrioventricular (tricuspid) valve
9. Right ventricle
10. Interventricular septum

Figure 12.17
Coronal section of the mammalian
(sheep) heart showing openings of
coronary arteries.
1. Opening of brachiocephalic artery
2. Opening of left coronary artery
3. Opening of right coronary artery
4. Aortic valve
5. Coronary vessel

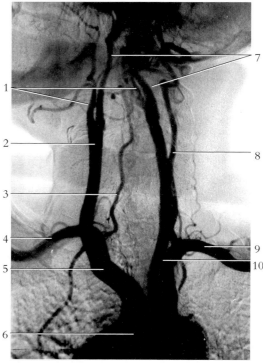

Figure 12.18

An angiogram showing the aortic arch and its branches.

1. External carotid arteries
2. Right common carotid artery
3. Right vertebral artery
4. Right subclavian artery
5. Brachiocephalic trunk
6. Aortic arch
7. Internal carotid arteries
8. Left vertebral artery
9. Left subclavian artery
10. Left common carotid artery

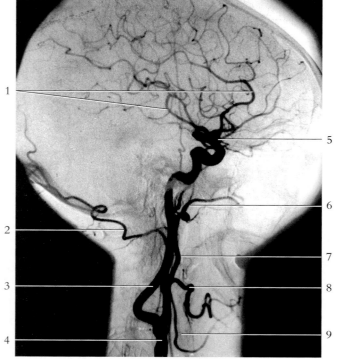

Figure 12.19

An angiogram showing the branches of the common carotid and external carotid arteries.

1. Meningeal arteries
2. Occipital artery
3. Internal carotid artery
4. Common carotid artery
5. Internal carotid artery to cerebral arterial circle (circle of Willis)
6. Maxillary artery
7. External carotid artery
8. Facial artery
9. Superior thyroid artery

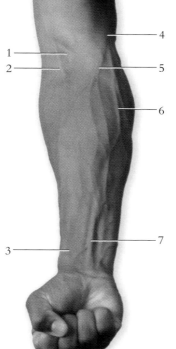

Figure 12.20

Surface anatomy identifying the superficial vessels of the forearm.

1. Cubital fossa
2. Cephalic vein
3. Radial artery (arterial pressure point)
4. Basilic vein
5. Median cubital vein
6. Median antebrachial vein
7. Tendon of palmaris longus m.

Figure 12.21

Arteries of the lower abdominal cavity.

1. Adrenal gland
2. Renal artery
3. Renal vein
4. Right kidney
5. Ureter
6. Inferior vena cava
7. Abdominal aorta
8. Right common iliac artery

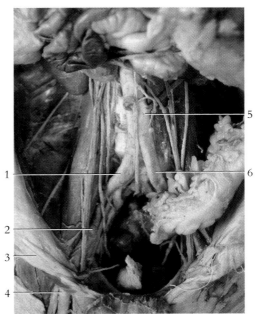

Figure 12.22
Arteries of the pelvic cavity.
1. Right common iliac artery
2. External iliac artery
3. Inguinal ligament
4. Femoral artery
5. Abdominal aorta
6. Left common iliac artery

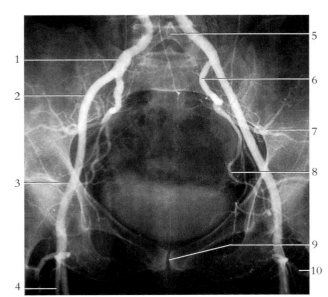

Figure 12.23
An angiogram of the common iliac arteries and their branches.
1. Common iliac artery
2. External iliac artery
3. Femoral artery
4. Deep femoral artery
5. Lumbar vertebra
6. Internal iliac artery
7. Gluteal arteries
8. Obturator artery
9. Symphysis pubis
10. Lateral circumflex femoral artery

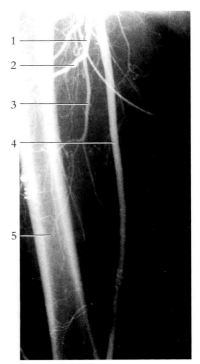

Figure 12.24
An angiogram of the arteries of the right thigh.
1. Deep femoral artery
2. Lateral circumflex femoral artery
3. Medial femoral circumflex artery
4. Femoral artery
5. Femur

Figure 12.25
Arterial plaque from femoral arteries.

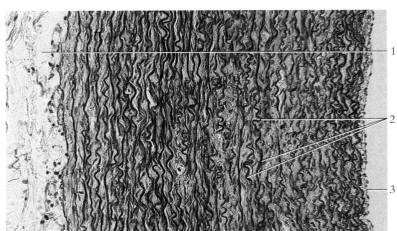

Figure 12.26   200X
Wall of elastic artery.
1. Tunica adventitia
2. Elastic laminae (in tunica media)
3. Tunica intima

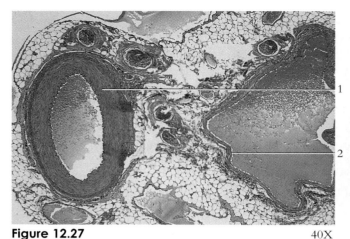

Figure 12.27 40X
Artery and vein.
1. Artery
2. Vein

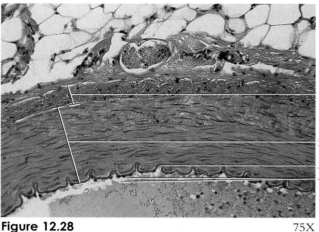

Figure 12.28 75X
Wall of muscular artery.
1. Tunica adventitia
2. Tunica media
3. Internal elastic membrane
4. Endothelial cells

Photo courtesy of Scott Miller

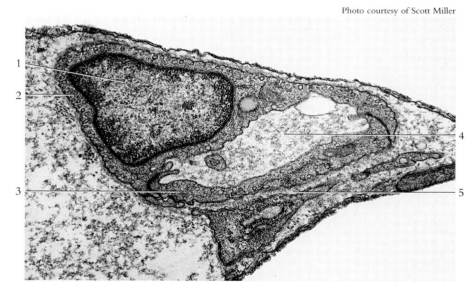

Figure 12.29
SEM photomicrograph of a capillary.
1. Nucleus
2. Endocytic vesicles
3. Endothelial cell
4. Lumen of capillary
5. Basal lamina

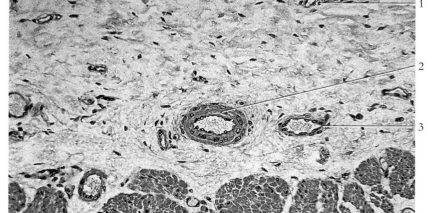

Figure 12.30
Arteriole, capillary, and venule.
1. Capillary
2. Arteriole
3. Venule

200X

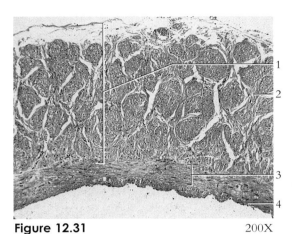

Figure 12.31   200X
Wall of large vein.
1. Tunica adventita
2. Longitudinally oriented
   smooth muscle
3. Tunica media
4. Tunica intima

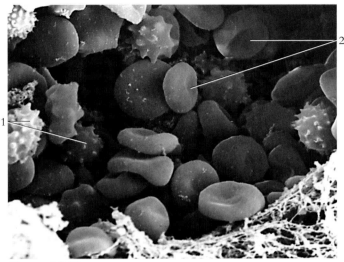

Figure 12.32
Electron micrograph of blood cells in the lumen of a blood vessel.
1. Leukocytes
2. Erythrocytes

Photo courtesy of Clifford E. Keeney

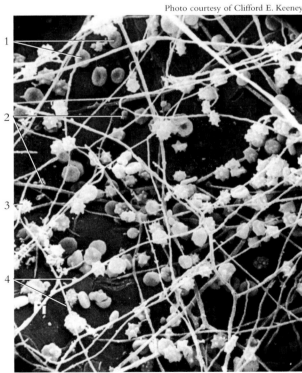

Figure 12.33
Electron micrograph of a blood clot.
1. Erythrocytes   3. Leukocyte
2. Thrombocytes   4. Fibrin strand

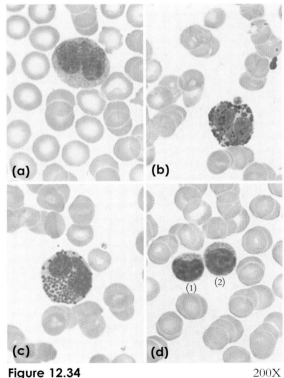

Figure 12.34   200X
Types of leukocytes.
(a) Neutrophil   (d) Lymphocyte (1)
(b) Basophil   Monocyte (2)
(c) Eosinophil

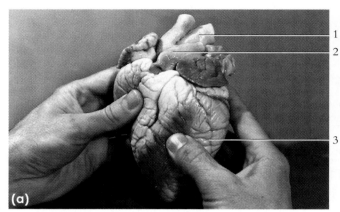

(a)

Position the heart so the ventral surface faces you. Notice the thicker ventricular walls, especially the left ventricle.
1. Aortic arch
2. Pulmonary trunk
3. Left ventricle

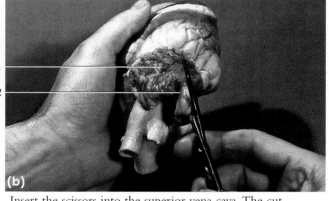

(b)

Insert the scissors into the superior vena cava. The cut should expose the interior of the right atrium. Notice the right atrioventricular (tricuspid) valve.
1. Right atrium
2. Superior vena cava

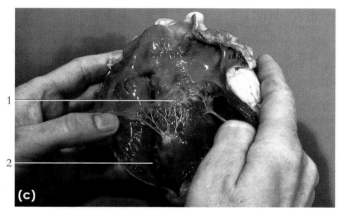

(c)

Continue the incision through the right ventricle to the apex of the heart. Observe the structure of the valve.
1. Right atrioventricular valve
2. Right ventricle

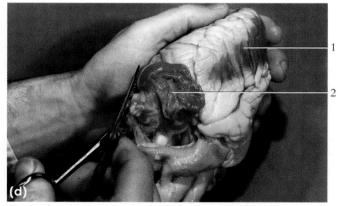

(d)

Begin the next incision in the left atrium. This time, continue through both the atrium and the ventricle.
1. Left ventricle
2. Left atrium

(e)

Expose the left ventricle and atrium. Notice the difference between the right and left ventricles, especially the thicker muscular wall of the left ventricle.
1. Left atrioventricular (bicuspid) valve.
2. Left ventricle

Figure 12.35
Sheep heart dissection (a-e).

The lymphatic system is closely interrelated to the circulatory system. The functions of the lymphatic system are basically fourfold: 1) it transports excess interstitial (tissue) fluid, which was initially formed as a blood filtrate, back to the bloodstream; 2) it maintains homeostasis around body cells by providing a constantly moist intracellular environment, which assists movements of materials into and out of cells; 3) it serves as the route by which absorbed fat from the small intestine is transported to the blood; and 4) it helps provide immunological defenses against disease-causing agents.

Lymph capillaries drain tissue fluid, which is formed from blood plasma; when this fluid enters lymph capillaries, it is called **lymph**. Lymph is returned to the venous system via two large lymph ducts—the **thoracic duct** and the **right lymphatic duct** (fig. 13.1). On the way to these drainage ducts, lymph filters through **lymph nodes**, which contain phagocytic cells and germinal centers that produce lymphocytes. The **spleen** and **thymus** are considered lymphoid organs because they also produce lymphocytes.

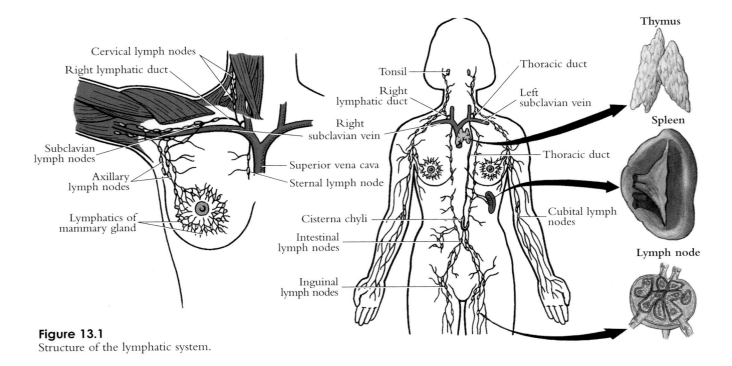

Figure 13.1
Structure of the lymphatic system.

Figure 13.2
Thymus.
1. Cortex
2. Medulla

7X

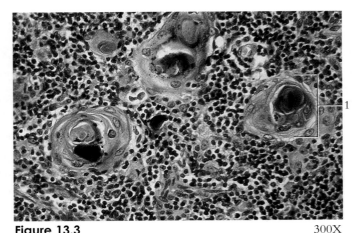

Figure 13.3
Thymic medulla.
1. Hassall's corpuscle

300X

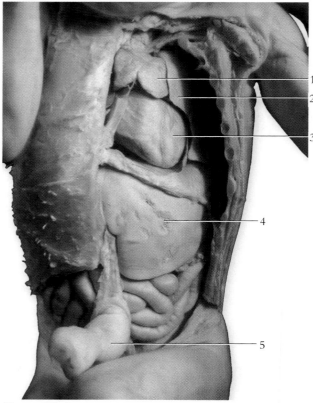

Figure 13.4
The thymus within a fetus, during
the third trimester of development.

1. Thymus
2. Lung
3. Heart
4. Liver
5. Umbilical cord

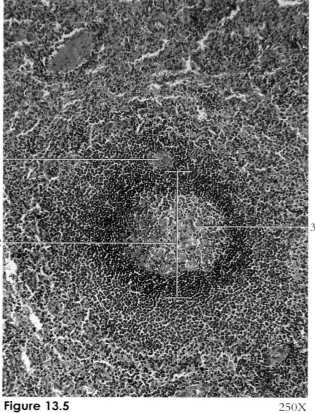

Figure 13.5   250X
Spleen.
  1. Central artery
  2. Splenic nodule
  3. Germinal center

Figure 13.6
The spleen and pancreas.
  1. Spleen
  2. Splenic artery
  3. Splenic vein
  4. Pancreas
  5. Pancreatic duct

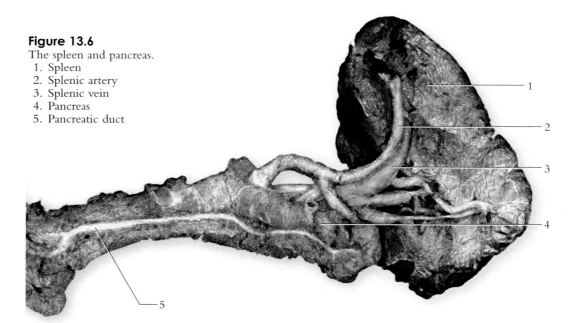

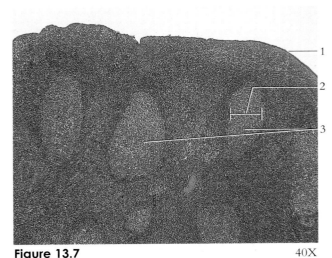

Figure 13.7   40X
Palatine tonsil.
1. Oral mucosa
2. Lymphatic nodule
3. Germinal centers

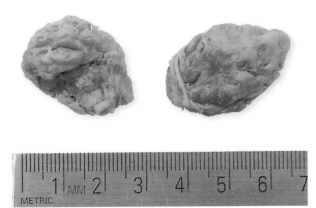

Figure 13.8
Palatine tonsils that have been removed in a tonsillectomy.
Chronic tonsillitis generally requires a tonsillectomy.

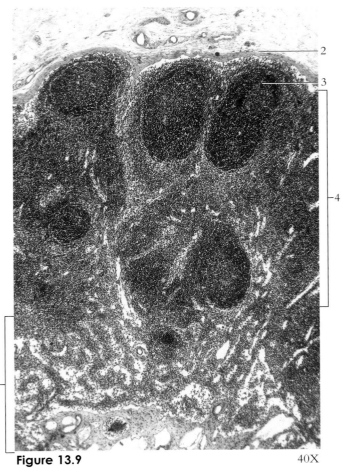

Figure 13.9   40X
Lymph node.
1. Medulla of lymph node   3. Lymphatic nodule
2. Capsule   4. Cortex of lymph node

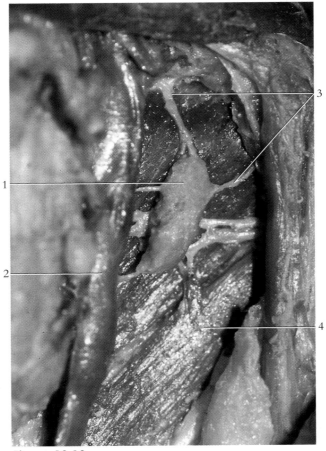

Figure 13.10
A lymph node.
1. Lymph node   3. Lymphatic vessels
2. Vein   4. Muscle

14 Respiratory System

The respiratory system is made up of organs and structures that function together to bring gases in contact with the blood of the circulatory system. This system consists of the **nasal cavity, pharynx, larynx**, and **trachea**, and the **bronchi, bronchioles**, and **pulmonary alveoli** within the lungs (fig. 14.2). The functions of the respiratory system are gas exchange, sound production, assistance in abdominal compression, and coughing and sneezing.

The nasal cavity has a bony and cartilaginous support. The ciliated, mucous lining of the upper respiratory tract warms, moistens, and cleanses inspired air. The **paranasal sinuses** are found in the maxillary, frontal, sphenoid, and ethmoid bones. The **pharynx** is an organ with a funnel-shaped passageway that connects the oral and nasal cavities with the larynx. The cartilaginous **larynx** keeps the passageway to the trachea open during breathing and closes the respiratory passageway during swallowing. It also contains the **vocal folds (vocal cords)**. The **trachea** is a rigid tube, supported by rings of cartilage, that leads from the larynx to the bronchial tree. **Pulmonary alveoli** are the functional units of the lungs where gas exchange occurs; they are small, numerous, thin-walled air sacs. The right and left **lungs** are separated by the **mediastinum**; each lung is divided into **lobes** and **lobules**.

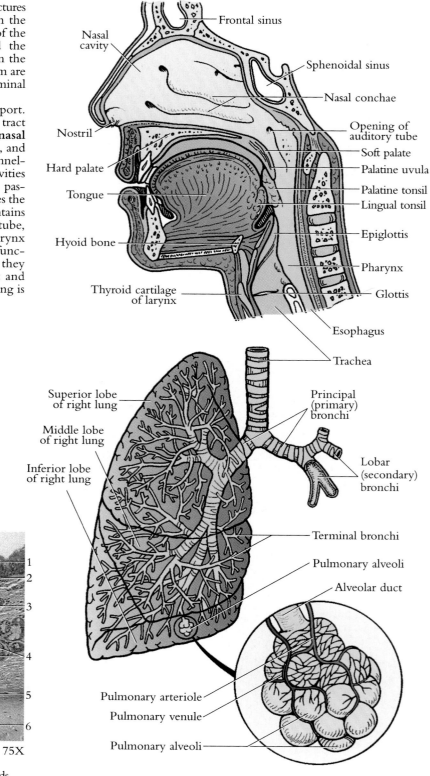

Figure 14.2

Structure of the respiratory system.

Figure 14.1   75X

Tracheal wall.

1. Respiratory epithelium
2. Basement membrane
3. Duct of seromucous gland
4. Seromucous glands
5. Perichondrium
6. Hyaline cartilage

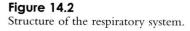

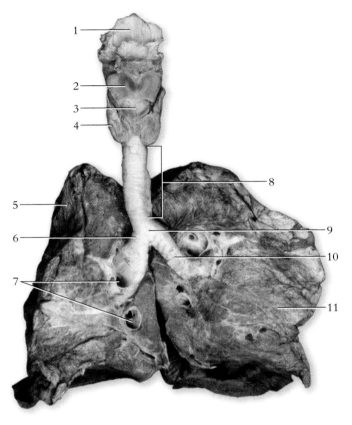

Figure 14.3
An anterior view of the larynx, trachea, and lungs.

1. Epiglottis
2. Thyroid cartilage
3. Cricoid cartilage
4. Thyroid gland
5. Right lung
6. Right principal (primary) bronchus

7. Pulmonary vessels
8. Trachea
9. Carnia
10. Left principal (primary) bronchus
11. Left lung

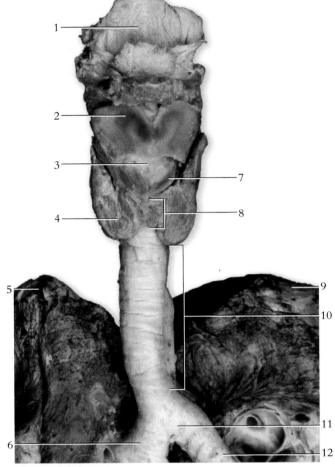

Figure 14.4
An anterior view of the larynx and trachea.

1. Epiglottis
2. Thyroid cartilage
3. Cricoid cartilage
4. Thyroid gland
5. Superior lobe of right lung
6. Right principal (primary) bronchus

7. Cricothyroid ligament
8. Isthmus of thyroid gland
9. Superior lobe of left lung
10. Trachea
11. Carnia
12. Left principal (primary) bronchus

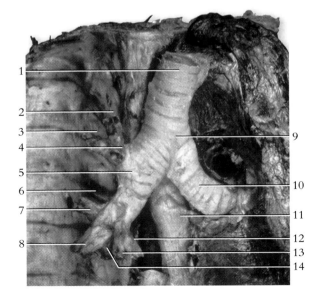

Figure 14.5
An anterior view of bronchi.

1. Trachea
2. Apical segmental bronchus
3. Posterior segmental bronchus
4. Anterior segmental bronchus
5. Right principal bronchus
6. Lateral segmental bronchus
7. Medial segmental bronchus

8. Anterior basal segmental bronchus
9. Carina
10. Left principal bronchus
11. Esophagus
12. Medial basal segmental bronchus
13. Posterior basal segmental bronchus
14. Lateral segmental bronchus

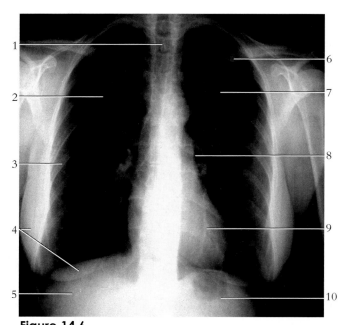

Figure 14.6

Radiograph of the thorax.

1. Thoracic vertebra
2. Right lung
3. Rib
4. Image of right breast
5. Diaphragm/liver
6. Clavicle
7. Left lung
8. Mediastinum
9. Heart
10. Diaphragm/stomach

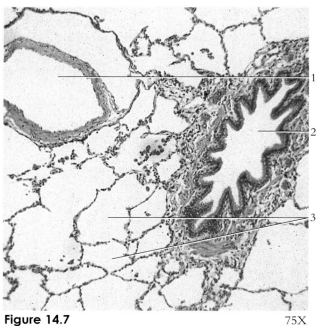

Figure 14.7   75X

Bronchiole.

1. Pulmonary arteriole
2. Bronchiole
3. Pulmonary alveoli

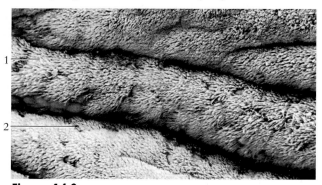

Figure 14.8

Electron micrograph of the lining of the trachea.

1. Cilia   2. Goblet cell

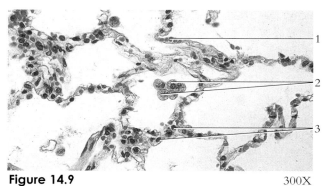

Figure 14.9   300X

Pulmonary alveoli.

1. Capillary in alveolar wall
2. Macrophages
3. Type II pneumocytes

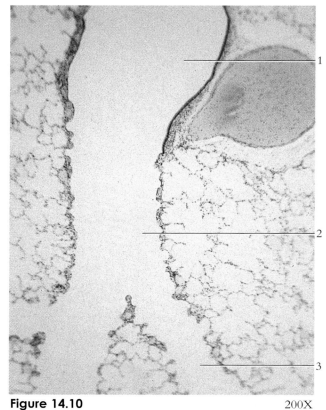

Figure 14.10   200X

Terminal bronchiole.

1. Terminal bronchiole
2. Respiratory bronchiole
3. Alveolar duct

The digestive system consists of a **gastrointestinal tract (GI tract)** and **accessory digestive organs**. Most of the food we eat is not suitable for cellular utilization until it is mechanically and chemically reduced to forms that can be absorbed through the intestinal wall and transported to the cells by the blood or lymph. Ingested food is not technically in the body until it is absorbed and, in fact, a large portion of consumed food is not digested at all but rather passes through as fecal material.

The functions of the principal regions and organs of the digestive system are presented in table 15.1. The digestive system is diagrammed in figure 15.1.

Table 15.1 Regions and structures of the digestive system.

Region or Structure	Function
Gastrointestinal tract	
Oral cavity	Ingests food; receives saliva and initiates digestion of carbohydrates; mastication (chewing); forms bolus (food mass); deglutition (swallowing)
Pharynx	Receives bolus from oral cavity and passes it to esophagus
Esophagus	Transports bolus to stomach by peristalsis
Stomach	Receives bolus from esophagus; forms chyme (paste-like food) initiates digestion of proteins; moves chyme into duodenum; participates in vomiting
Small intestine	Receives chyme from stomach, along with secretions from liver and pancreas; chemically and mechanically breaks down chyme; absorbs nutrients; transports wastes to large intestine
Large intestine	Receives undigested wastes from small intestine; absorbs water and electrolytes; forms and stores feces, and expels them through defecation
Accessory digestive organs	
Teeth	Mechanically pulverize food
Tongue	Manipulate food and assists in swallowing
Salivary glands	Secrete saliva which aids in formation of bolus; initiates digestion of carbohydrates
Liver	Production of bile; storage of iron and copper; conversion of glucose to glycogen and storage of glycogen; synthesis of certain vitamins; production of urea; synthesis of fibrinogen and prothrombin used for clotting of blood; phagocytosis of foreign material in blood; detoxifies harmful substances in body; storage of blood cells; hemopoiesis in fetus and newborn
Gallbladder	Concentration and storage of bile necessary for emulsification of fats; releases bile into duodenum as pyloric sphincter of stomach opens
Pancreas	Production and secretion of pancreatic juice containing digestive enzymes; production and secretion of the hormones insulin and glucagon

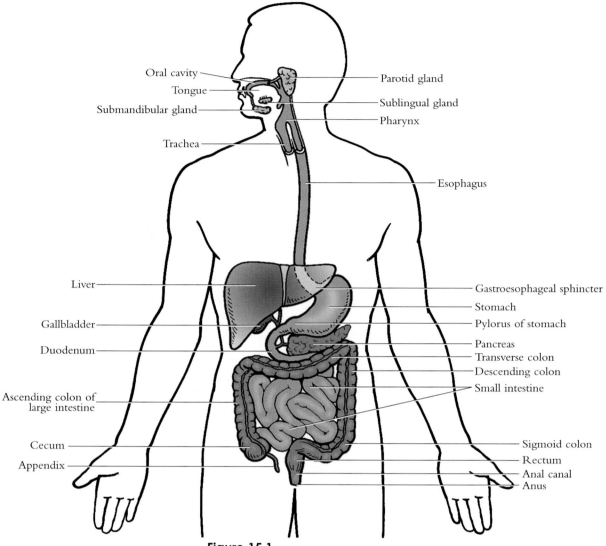

Figure 15.1
Structure of the digestive system.

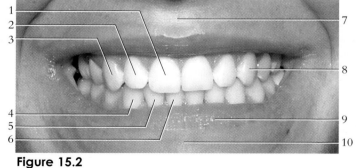

Figure 15.2
Oral region, lips, and teeth.

1. Medial incisor
2. Lateral incisor
3. Canine
4. Canine
5. Lateral incisor
6. Medial incisor
7. Philtrum
8. Canine
9. Inferior lip
10. Mentolabial sulcus

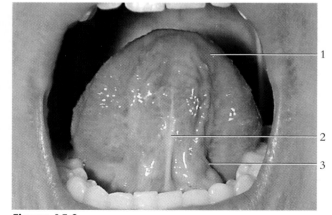

Figure 15.3
Structures of the oral cavity with the mouth
open and the tongue elevated.

1. Tongue
2. Lingual frenulum
3. Opening of
   submandibular duct

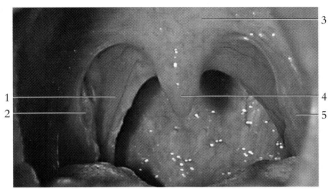

Figure 15.4
Superficial structures of the oral cavity.
1. Pharyngopalatine arch
2. Palatine tonsil
3. Soft palate
4. Palatine uvula
5. Glossopalatine arch

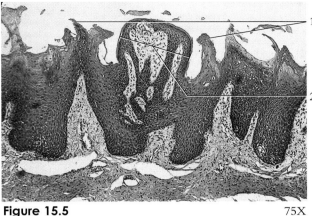

Figure 15.5 75X
Filiform and fungiform papillae.
1. Filiform papillae
2. Fungiform papilla

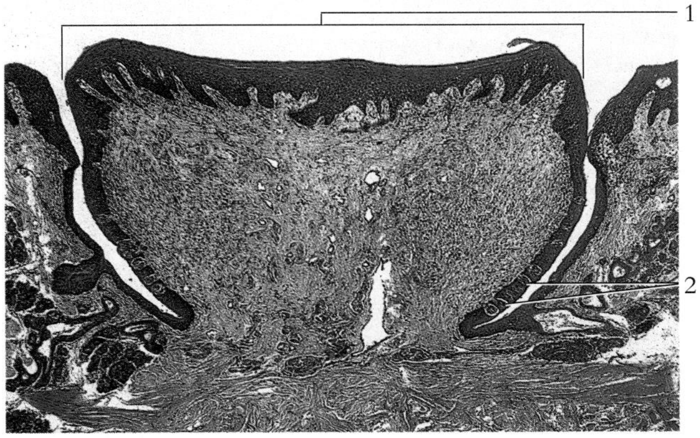

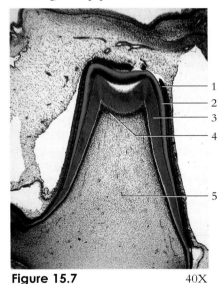

Figure 15.6 40X
Vallate papilla
1. Vallate papilla
2. Taste buds

Figure 15.7 40X
Developing tooth.
1. Ameloblasts
2. Enamel
3. Dentin
4. Odontoblasts
5. Pulp

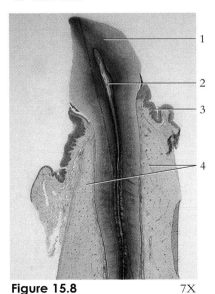

Figure 15.8 7X
Mature tooth.
1. Dentin (enamel has been dissolved away)
2. Pulp
3. Gingiva
4. Alveolar bone

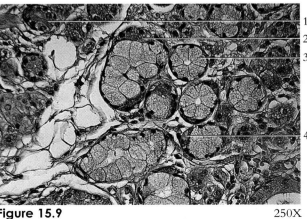

Figure 15.9 250X
Acini of salivary tissue.
1. Serous acinus
2. Serous demilune on mucous acinus
3. Mucous acinus
4. Serous demilune on mucous acinus

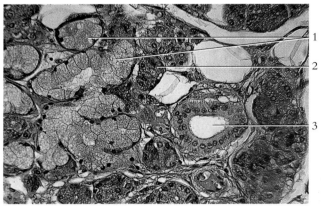

Figure 15.10   250X
Striated duct.
1. Mucous acini   3. Lumen of striated duct
2. Serous acinus

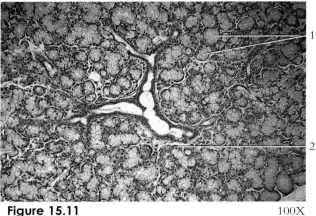

Figure 15.11   100X
Sublingual gland (mostly mucous, some serous).
1. Mucous acini
2. Serous demilune

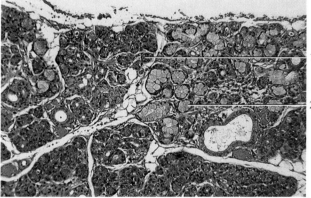

Figure 15.12   100X
Submandibular gland (about one-half mucous and one-half serous).
1. Serous acinus
2. Mucous acinus

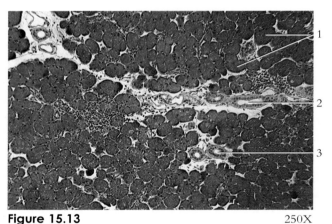

Figure 15.13   250X
Parotid gland (purely serous).
1. Serous acini   3. Lumen of striated duct
2. Lumen of excretory duct

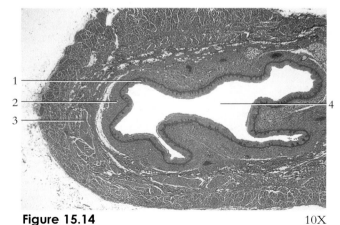

Figure 15.14   10X
Cross section of esophagus.
1. Mucosa   3. Muscularis
2. Submucosa   4. Lumen

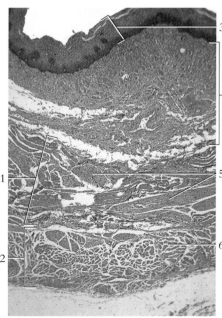
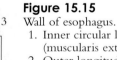

Figure 15.15
Wall of esophagus.
1. Inner circular layer (muscularis externa)
2. Outer longitudinal layer (muscularis externa)
3. Mucosa
4. Submucosa
5. Smooth muscle
6. Skeletal muscle

30X

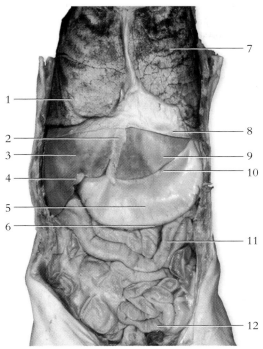

Figure 15.16
An anterior aspect of the trunk.
1. Right lung
2. Falciform ligament
3. Right lobe of liver
4. Gallbladder
5. Body of stomach
6. Greater curvature of stomach
7. Left lung (reflected)
8. Diaphragm
9. Left lobe of liver
10. Lesser curvature of stomach
11. Transverse colon
12. Small intestine

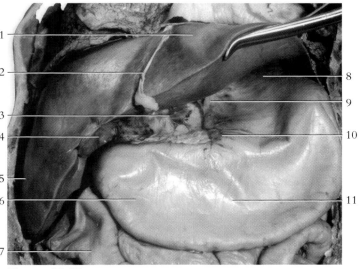

Figure 15.17
An anterior view of the stomach and liver.
1. Left lobe of liver
2. Falciform ligament
3. Celiac trunk (covered by hepatoduodenal ligament)
4. Gallbladder
5. Right lobe of liver
6. Pylorus of stomach
7. Transverse colon
8. Fundus of stomach
9. Lesser omentum
10. Gastric vein (traversing through omentum
11. Body of stomach

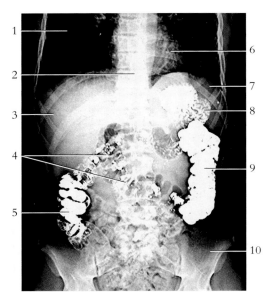

Figure 15.18
A CT scanogram of the trunk.
1. Right lung surrounded by pleural cavity
2. Thoracic vertebra
3. Liver
4. Small intestine
5. Ascending colon
6. Heart
7. Diaphragm
8. Stomach
9. Descending colon
10. Ilium

Figure 15.19
Thoracic and abdominal viscera.
1. Lungs
2. Heart (surrounded by pericardial fat)
3. Diaphragm
4. Liver
5. Cecum
6. Vagus nerves
7. Phrenic nerves
8. Small intestine

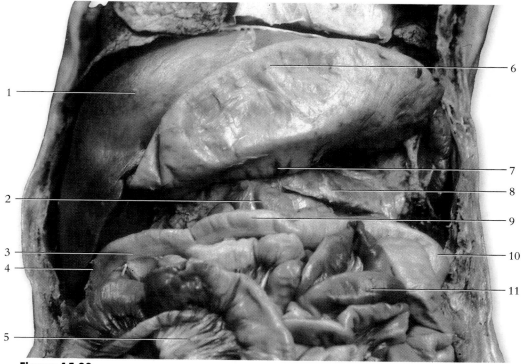

Figure 15.20

Upper abdominal organs. The greater omentum has been removed and the stomach reflected.

1. Right lobe of liver
2. Duodenum
3. Taeniae coli
4. Right colic (hepatic) flexure
5. Mesentery
6. Greater curvature of stomach
7. Lesser omentum
8. Pancreas
9. Transverse colon
10. Left colic (splenic) flexure
11. Small intestine

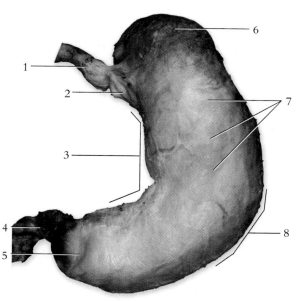

Figure 15.21

Major regions and structures of the stomach.

1. Esophagus
2. Cardiac portion of stomach
3. Lesser curvature of stomach
4. Duodenum
5. Pylorus of stomach
6. Fundus of stomach
7. Body of stomach
8. Greater curvature of stomach

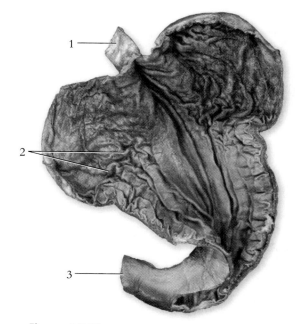

Figure 15.22

Interior aspect of the stomach.

1. Esophagus
2. Gastric folds (rugae)
3. Duodenum

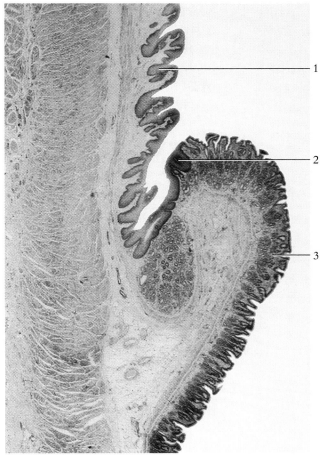

Figure 15.23   10X

Junction of esophagus and stomach.

1. Epithelium of esophagus
2. Abrupt change in epithelium
3. Epithelium of stomach

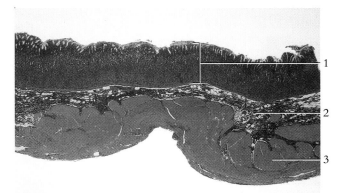

Figure 15.24   10X

Wall of stomach.

1. Mucosa
2. Submucosa
3. Muscularis externa

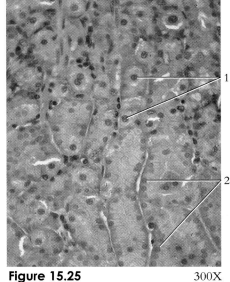

Figure 15.25   300X

Gastric glands.

1. Parietal cells
2. Chief cells

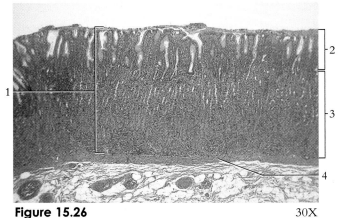

Figure 15.26   30X

Mucosa of stomach body.

1. Mucosa
2. Gastric pits
3. Gastric glands
4. Muscularis mucosae

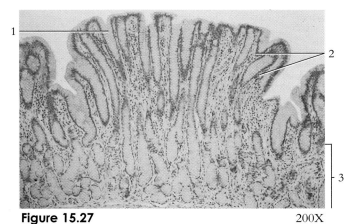

Figure 15.27   200X

Mucosa of stomach pylorus.

1. Gastric pit
2. Lamina propria
3. Gastric glands

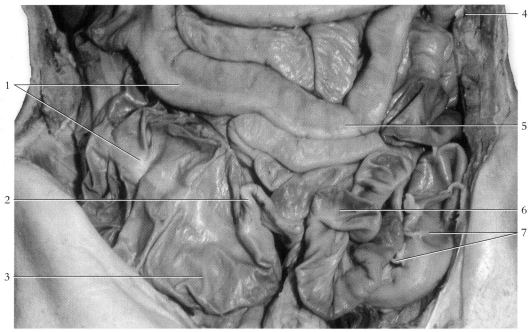

Figure 15.28
An anterior aspect of the small and large intestines with associated structures.

1. Taeniae coli
2. Appendix
3. Cecum
4. Left colic (splenic) flexure
5. Transverse colon
6. Small intestine
7. Epiploic appendages

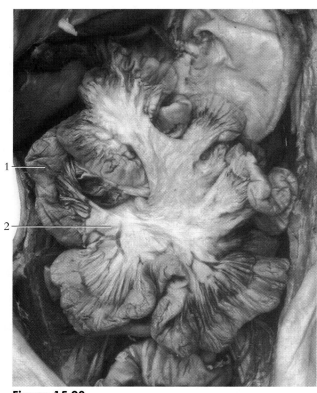

Figure 15.29
Mesentery and small intestine.
1. Serosa (visceral peritoneum)
2. Mesentery

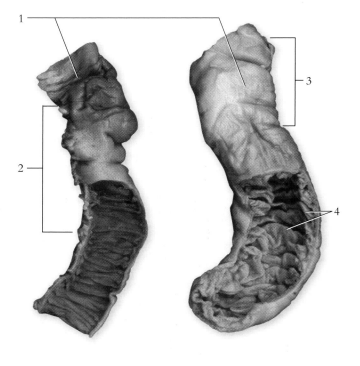

Figure 15.30
Sections of the small intestine.
1. Serosa (visceral peritoneum)
2. Ileum
3. Jejunum
4. Plicae circulares

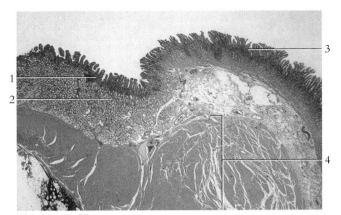

Figure 15.31 10X
Junction of pylorus and duodenum.
1. Duodenal epithelium
2. Duodenal (Brunner's) glands
3. Stomach epithelium
4. Pyloric sphincter

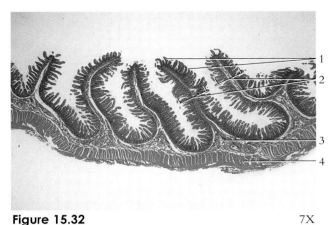

Figure 15.32 7X
Small intestine.
1. Intestinal villi
2. Plicae circulares
3. Submucosa
4. Muscularis externa

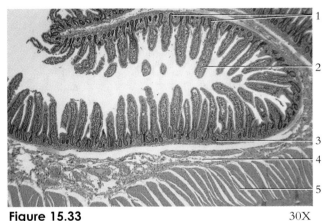

Figure 15.33 30X
Small intestine.
1. Intestinal glands
2. Intestinal villi
3. Muscularis mucosae
4. Submucosa
5. Muscularis externa

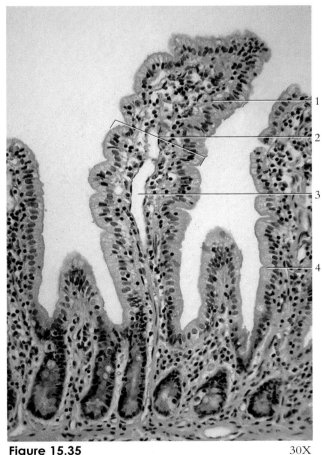

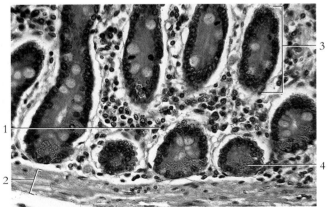

Figure 15.34 300X
Intestinal glands.
1. Eosinophil within lamina propria
2. Muscularis mucosae
3. Intestinal gland
4. Paneth cells

Figure 15.35 30X
Intestinal villus.
1. Lamina propria
2. Intestinal villus
3. Central lacteal
4. Mucosa of simple columnar epithelium

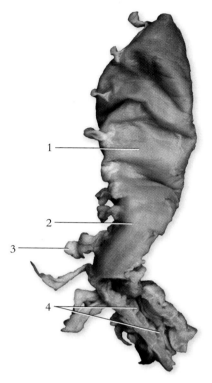

Figure 15.36
Section of the large intestine (colon).
1. Haustrum
2. Taeniae coli
3. Epiploic appendage
4. Semilunar folds
   (plicae) of colon

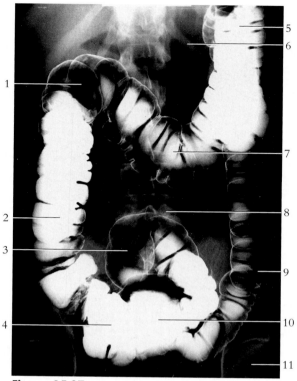

Figure 15.37
Radiograph of the large intestine.
1. Right colon (hepatic) flexure
2. Ascending colon
3. Sigmoid colon
4. Cecum
5. Left colic (splenic) flexure
6. 12th rib
7. Transverse colon
8. Lumbar vertebra
9. Descending colon
10. Rectum
11. Hip joint

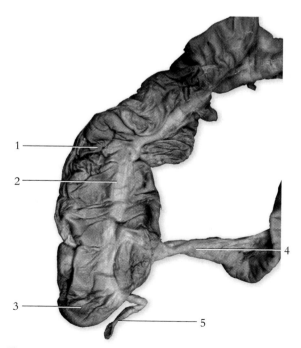

Figure 15.38
The cecum and appendix.
1. Ascending colon
2. Taenia coli
3. Cecum
4. Ileum
5. Appendix

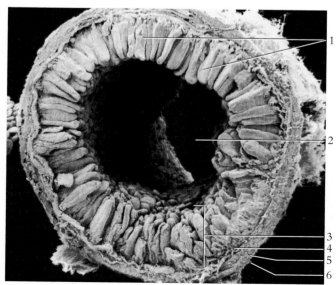

Figure 15.39
Electron micrograph of the ileum, shown in cross section.
1. Intestinal villi
2. Lumen
3. Mucosa
4. Submucosa
5. Tunica muscularis
6. Adventitia

Figure 15.40
An inflamed appendix that has been removed in an appendectomy. The chief danger of appendicitis is the the appendix might rupture and produce peritonitis.

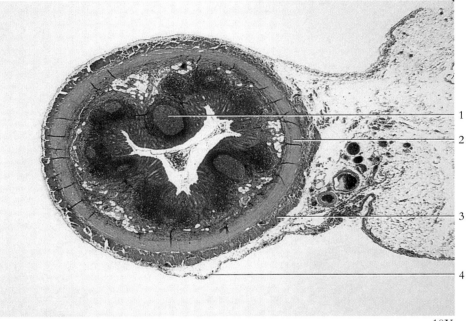

10X

Figure 15.41
Appendix, shown in cross section.
1. Lymphatic nodule
2. Circular layer of muscularis externa
3. Longitudinal layer of muscularis externa
4. Serosa (visceral peritoneum)

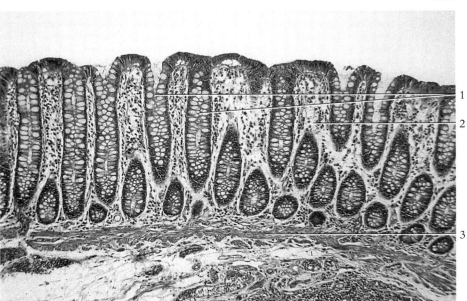

75X

Figure 15.42
Large intestine.
1. Glands
2. Lamina propria
3. Muscularis mucosae

Figure 15.43
An inferior view of the liver and gallbladder.
1. Gallbladder
2. Right lobe of liver
3. Cystic duct
4. Hepatic duct
5. Common bile duct
6. Falciform ligament
7. Left lobe of liver
8. Hepatic artery
9. Caudate lobe of liver
10. Hepatic portal vein

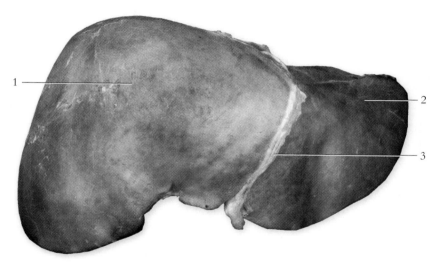

Figure 15.44
An anterior view of the liver.
1. Right lobe of liver
2. Left lobe of liver
3. Falciform ligament

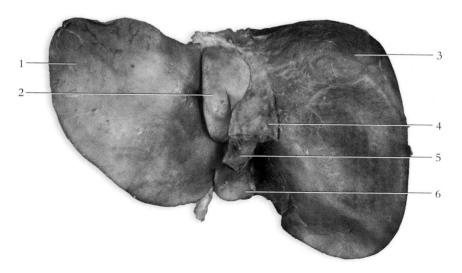

Figure 15.45
An inferior view of the liver.
1. Left lobe of liver
2. Caudate lobe of liver
3. Right lobe of liver
4. Hepatic portal vein
5. Hepatic artery
6. Quadrate lobe of liver

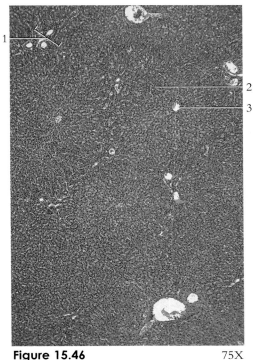

Figure 15.46   75X
Hepatic lobules.
1. Portal area   3. Central vein
2. Hepatic lobule

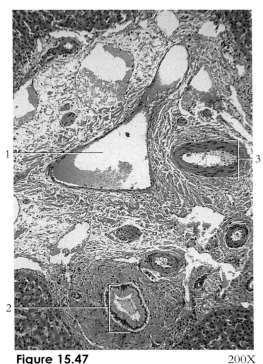

Figure 15.47   200X
Portal area.
1. Lumen of   2. Bile duct
   portal vein   3. Hepatic artery

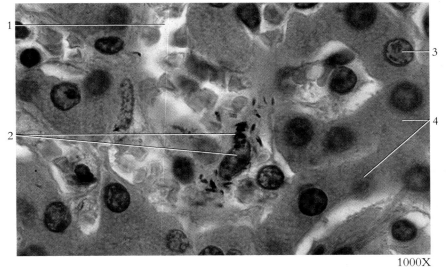

1000X

Figure 15.48
Hepatic plates.
1. Sinusoid containing red blood cells
2. Kupffer cell with debris in it's cytoplasm
3. Hepatocyte nucleus
4. Plate of hepatocytes

Figure 15.49
A gallbladder that has been removed
(cholecystectomy) and cut open to
remove it's gallstone (biliary calculus).
1. Gallbladder
2. Gallstone

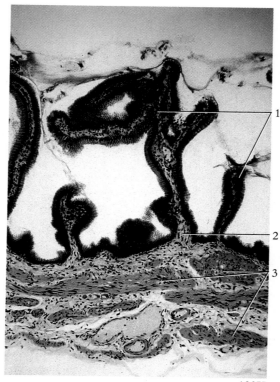

Figure 15.50
Gallbladder.
1. Mucosal folds
2. Lamina propria
3. Muscularis

100X

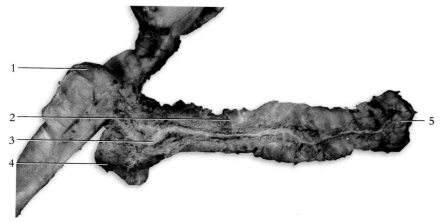

Figure 15.51
An anterior aspect of the
pancreas and pancreatic duct.
1. Duodenum
2. Body of pancreas
3. Pancreatic duct
4. Head of pancreas
5. Tail of pancreas

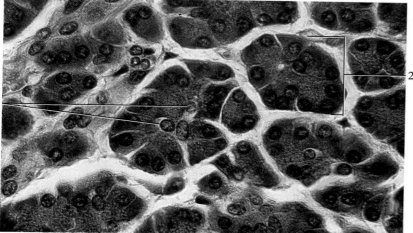

Figure 15.52
Pancreatic acini.
1. Centroacinar cells
2. Acinus

625X

The urinary system consists of the **kidneys, ureters, urinary bladder**, and **urethra** (fig. 16.1). The urinary system: 1) removes metabolic wastes from the blood and excretes it (as urine) to the outside during micturition (the physiological aspect of urination); 2) regulates, in part, the rate of red blood cell formation by secretion of the hormone **erythropoietin;** 3) assists the regulation of blood pressure by secreting the enzyme **renin;** 4) assists the regulation of calcium by activating vitamin D; and 5) assists the regulation of the volume, composition, and pH of body fluids.

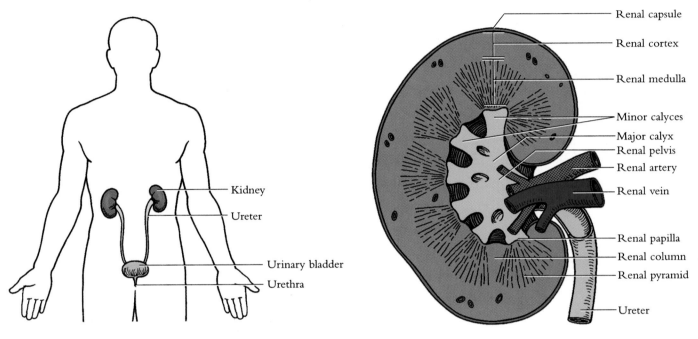

Figure 16.1
Organs of the urinary system.

Kidney
Ureter
Urinary bladder
Urethra

Figure 16.2
Structure of the kidney.

Renal capsule
Renal cortex
Renal medulla
Minor calyces
Major calyx
Renal pelvis
Renal artery
Renal vein
Renal papilla
Renal column
Renal pyramid
Ureter

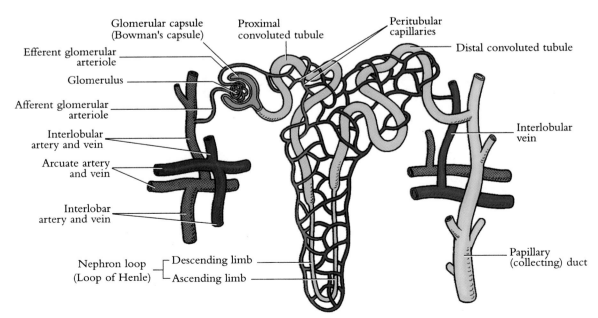

Glomerular capsule (Bowman's capsule)
Proximal convoluted tubule
Peritubular capillaries
Efferent glomerular arteriole
Distal convoluted tubule
Glomerulus
Afferent glomerular arteriole
Interlobular artery and vein
Interlobular vein
Arcuate artery and vein
Interlobar artery and vein
Nephron loop (Loop of Henle)
Descending limb
Ascending limb
Papillary (collecting) duct

Figure 16.3
Structure of the nephron.

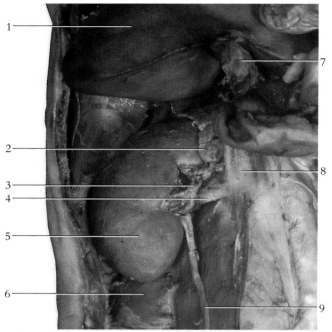

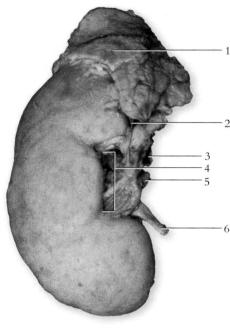

Figure 16.4

Kidney and ureter with overlying viscera removed.

1. Liver
2. Adrenal gland
3. Renal artery
4. Renal vein
5. Right kidney
6. Quadratus lumborum muscle
7. Gallbladder
8. Inferior vena cava
9. Ureter

Figure 16.5

An anterior view of the right kidney.

1. Adrenal gland
2. Suprarenal artery
3. Renal artery
4. Hilum
5. Renal vein
6. Ureter

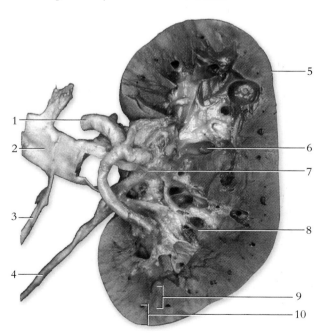

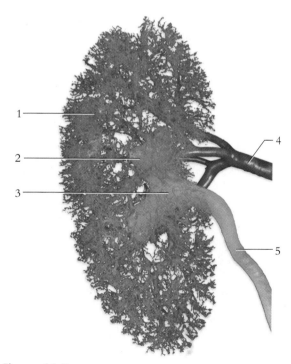

Figure 16.6

A coronal section of the left kidney.

1. Renal artery
2. Renal vein
3. Left testicular vein
4. Ureter
5. Renal capsule
6. Major calyx
7. Renal pelvis
8. Renal papilla
9. Renal medulla
10. Renal cortex

Figure 16.7

A plastic vascular cast of a kidney and ureter.

1. Arteriole network
2. Major calyx
3. Renal pelvis
4. Renal artery
5. Ureter

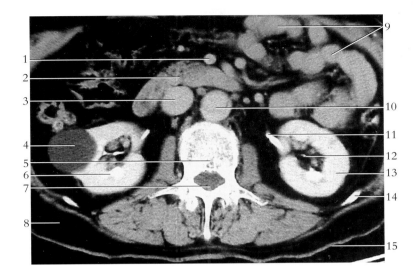

Figure 16.8
A transaxial CT image of abdominal viscera.
Note the large renal cyst in the right
kidney.
1. Superior mesenteric artery
2. Pancreas
3. Inferior vena cava
4. Renal crest
5. Body of lumbar vertebra
6. Right kidney
7. Vertebral foramen
8. Subcutaneous fat
9. Small intestine
10. Abdominal portion of aorta
11. Left ureter
12. Renal pelvis
13. Renal cortex
14. Rib
15. Skin

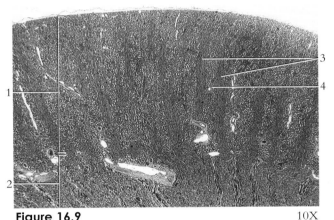

Figure 16.9   10X
Kidney.
1. Renal cortex   3. Medullary rays
2. Renal medulla   4. Cortical labyrinth

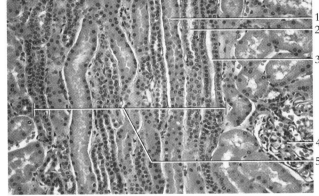

Figure 16.10   100X
Medullary ray.
1. Proximal tubule (of straight portion)   4. Glomerulus
2. Distal tubule (of straight portion)   5. Medullary ray
3. Collecting tubule

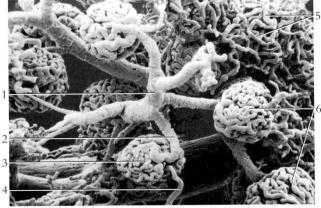

Figure 16.11
Electron micrograph of the renal cortex.
1. Interlobular artery   4. Efferent glomerular
2. Afferent glomerular   arteriole
   arteriole   5. Peritubular capillaries
3. Glomerulus   6. Glomeruli

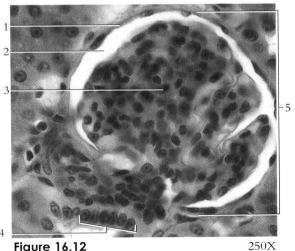

Figure 16.12   250X
Renal corpuscle.
1. Glomerular capsule   4. Macula densa of
2. Urinary space   distal tubule
3. Glomerulus   5. Renal corpuscle

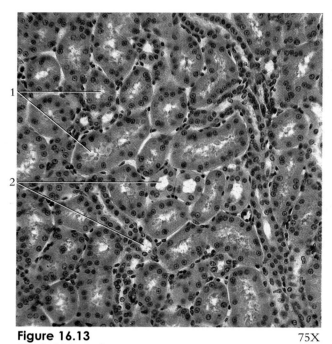

Figure 16.13   75X
Cortical labyrinth.
  1. Proximal convoluted tubule
  2. Distal convoluted tubule

Figure 16.14   100X
Renal papilla.
  1. Renal papilla
  2. Minor calyx
  3. Transitional epithelium

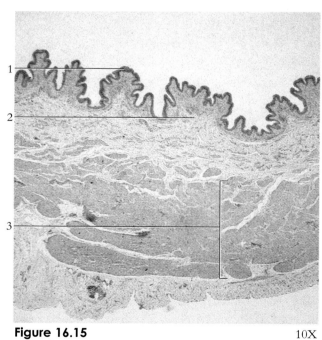

Figure 16.15   10X
Wall of urinary bladder.
  1. Transitional epithelium
  2. Lamina propria
  3. Muscularis

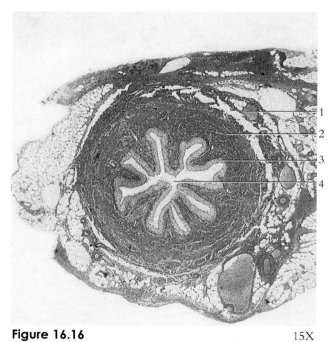

Figure 16.16   15X
Ureter, as shown in cross section.
  1. Adventitia   3. Mucosa
  2. Muscularis   4. Lumen

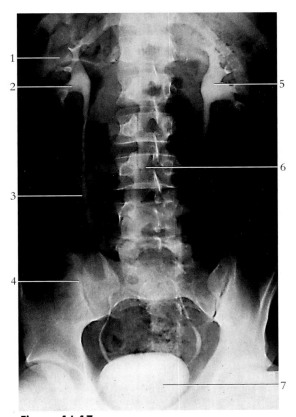

Figure 16.17
A pyelogram of the urinary system.
1. 12th rib
2. Major calyx
3. Ureter
4. Sacroiliac joint
5. Renal pelvis
6. 3rd lumbar vertebra
7. Urinary bladder

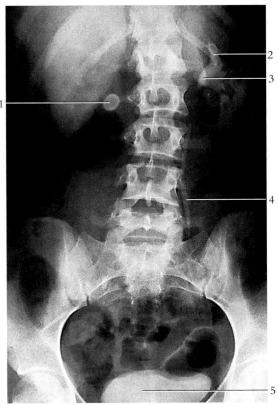

Figure 16.18
A radiograph showing a renal calculus (kidney stone) in the renal pelvis of the right kidney.
1. Renal calculus
2. Major calyx of left kidney
3. Renal pelvis of kidney
4. Left ureter
5. Urinary bladder

Figure 16.19
Renal calculi (stones) from the urinary bladders of two different patients. Renal calculi vary considerably in appearance and size and may develop in the renal pelvis of the kidney or in the urinary bladder.

17 Reproductive System

The organs of the male and female reproductive systems produce gametes and provide the mechanism for the union of these sex cells during the process of coitus (sexual intercourse). It is through sexual reproduction that individuals of a species are propagated, each having a genetic diversity inherited from both parents.

The male reproductive system consists of the **testes** (in the **scrotum**), excretory glands (**seminal vesicles, prostate**, and **bulbourethral glands**), and **penis**. The functions of the male reproductive system are to secrete sex hormones, produce spermatozoa, and ejaculate semen (spermatozoa and additives) into the vagina of the female.

The female reproductive system consists of the **ovaries, uterine tubes, vagina, external genitalia**, and **mammary glands**. The functions of the female reproductive system are to secrete sex hormones, produce ova, receive ejaculated semen from the erect penis of the male, nourish the developing embryo and fetus, deliver the baby, and nurse the infant once it is born.

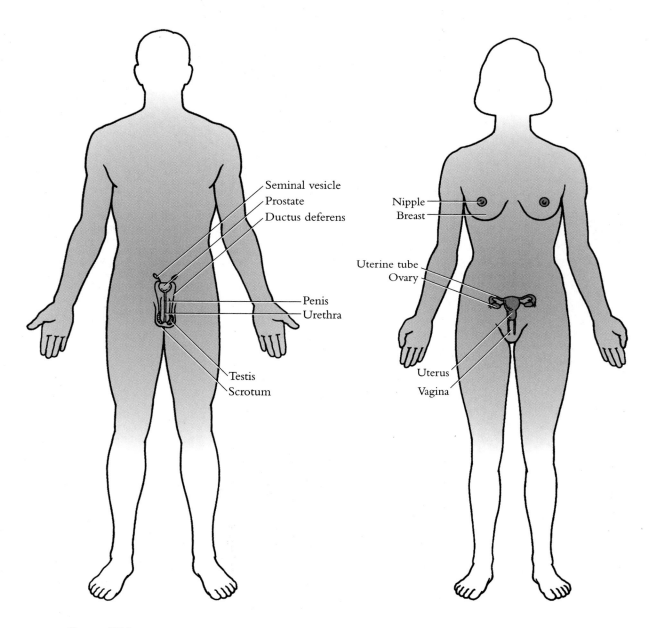

Figure 17.1
Organs of the male reproductive system.

Figure 17.2
Organs of the female reproductive system.

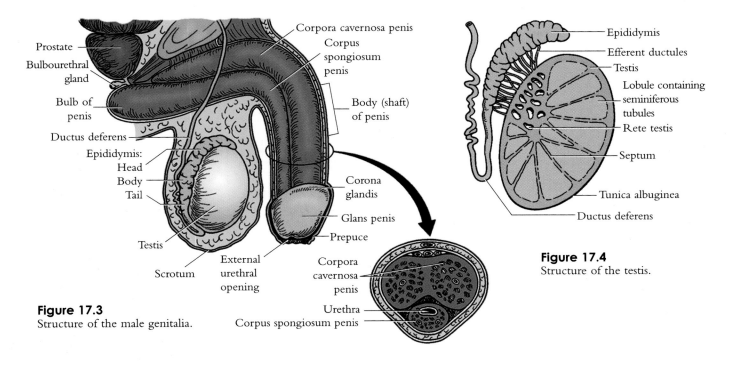

Figure 17.3
Structure of the male genitalia.

Prostate

Bulbourethral gland

Bulb of penis

Ductus deferens

Epididymis:
Head
Body
Tail

Testis

Scrotum

External urethral opening

Corpora cavernosa penis

Corpus spongiosum penis

Body (shaft) of penis

Corona glandis

Glans penis

Prepuce

Corpora cavernosa penis

Urethra

Corpus spongiosum penis

Figure 17.4
Structure of the testis.

Epididymis

Efferent ductules

Testis

Lobule containing seminiferous tubules

Rete testis

Septum

Tunica albuginea

Ductus deferens

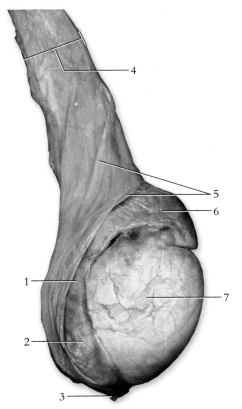

Figure 17.5
Testis and associated structures.

1. Body of epididymis
2. Tail of epididymis
3. Gubernaculum
4. Spermatic cord
5. Spermatic fascia
6. Head of epididymis
7. Testis

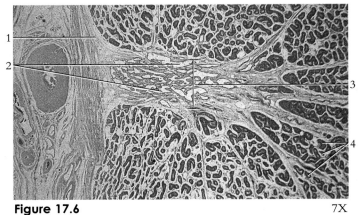

Figure 17.6   7X
Testis.

1. Tunic albuginea   3. Mediastinum
2. Tubules of rete testis   4. Seminiferous tubules

Figure 17.7
Cancerous testis; orchiectomy is the removal of a testis.

Figure 17.8
Electron micrograph of a seminiferous tubule.
1. Spermatozoa   3. Spermatogonia
2. Spermatids

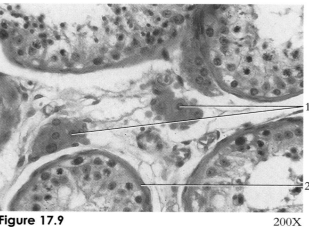

Figure 17.9   200X
Testis.
1. Interstitial (Leydig) cells
2. Seminiferous tubule

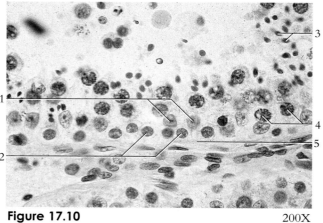

Figure 17.10   200X
Wall of seminiferous tubule.
1. Sustentacular (Sertoli) cells   4. Primary spermatocytes
2. Spermatogonia   5. Boundary of seminiferous
3. Spermatids   tubule

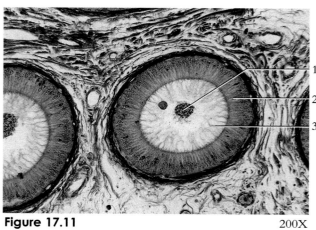

Figure 17.11   200X
Epididymis.
1. Sperm in lumen   3. Stereocilia
2. Pseudostratified columnar
    epithelium

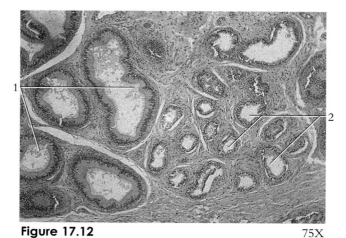

Figure 17.12   75X
Efferent ductules.
1. Duct of epididymis
2. Efferent ductules

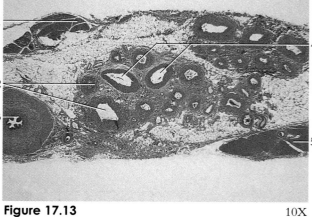

Figure 17.13   10X
Spermatic cord.
1. Cremaster muscle   3. Ductus deferens
2. Veins of the pampiniform   4. Testicular arterioles
    plexus   5. Cremaster muscle

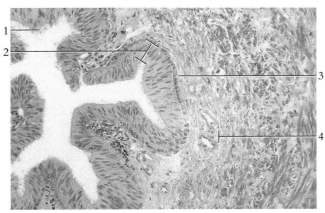

Figure 17.14 250X
Ductus deferens.
1. Stereocilia
2. Pseudostratified columnar epithelium
3. Lamina propria
4. Muscularis

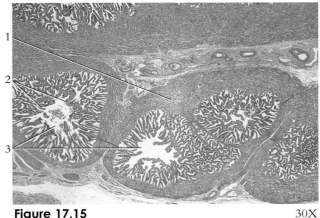

Figure 17.15 30X
Seminal vesicle.
1. Fibromuscular stroma
2. Mucosal folds
3. Lumen

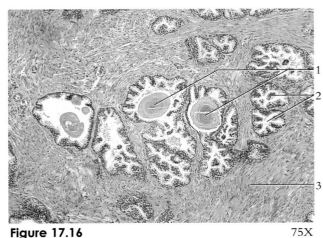

Figure 17.16 75X
Prostate.
1. Prostate concretions
2. Glandular acini
3. Fibromuscular stroma

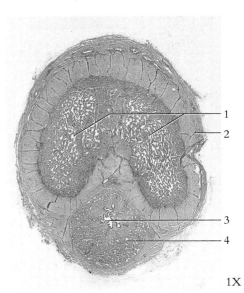

Figure 17.17
Penis.
1. Corpora cavernosa
2. Tunica albuginea
3. Urethra
4. Corpus spongiosum

1X

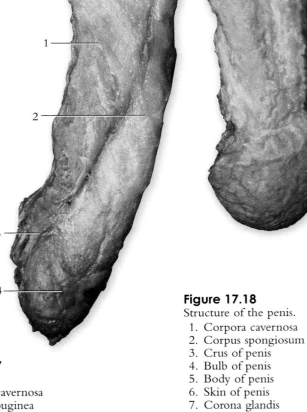

Figure 17.18
Structure of the penis.
1. Corpora cavernosa
2. Corpus spongiosum
3. Crus of penis
4. Bulb of penis
5. Body of penis
6. Skin of penis
7. Corona glandis
8. Glans penis

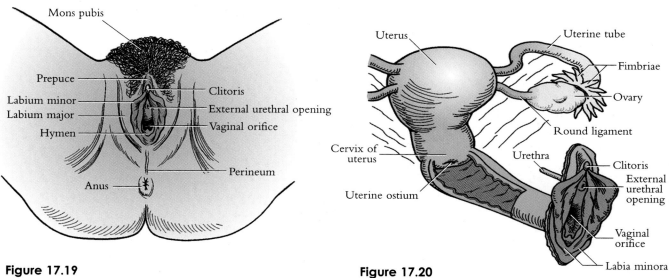

Figure 17.19
Female external genitalia.

Labels (Figure 17.19): Mons pubis, Prepuce, Labium minor, Labium major, Hymen, Anus, Clitoris, External urethral opening, Vaginal orifice, Perineum

Figure 17.20
External genitalia and internal reproductive organs of the female reproductive system.

Labels (Figure 17.20): Uterus, Uterine tube, Fimbriae, Ovary, Round ligament, Cervix of uterus, Uterine ostium, Urethra, Clitoris, External urethral opening, Vaginal orifice, Labia minora

Figure 17.21
External genitalia and vagina.
1. Mons pubis
2. Clitoris
3. Urethral opening
4. Labium minor
5. Labium major
6. Vagina (dissected open)
7. Fornix
8. Uterine ostium
9. Cervix of uterus

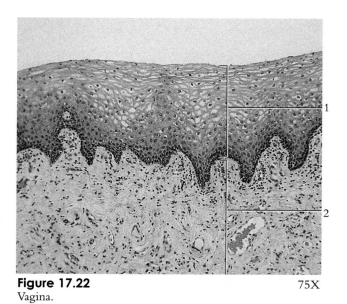

Figure 17.22   75X
Vagina.
1. Nonkeratinized stratified squamous epithelium
2. Lamina propria

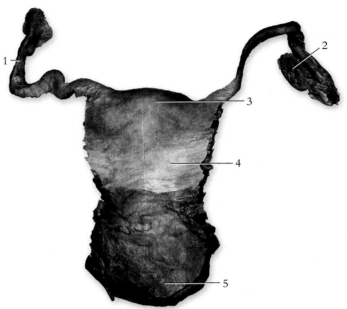

Figure 17.23

A posterior view of the uterus and uterine tubes.
1. Uterine tube
2. Fimbriae
3. Fundus of uterus
4. Body of uterus
5. Cervix of uterus

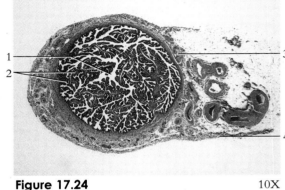

Figure 17.24   10X

Ampulla of uterine tube.
1. Lumen
2. Mucosal folds
3. Muscularis
4. Serosa

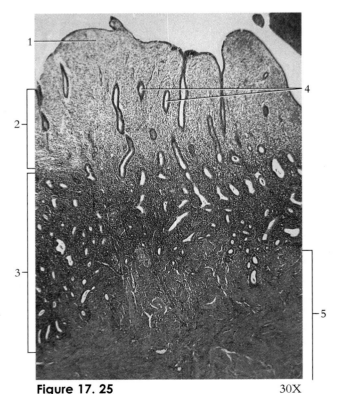

Figure 17. 25   30X

Uterus: proliferative phase.
1. Endometrial stroma
2. Lamina functionalis (endometrium)
3. Lamina basalis (endometrium)
4. Endometrial uterine glands
5. Myometrium

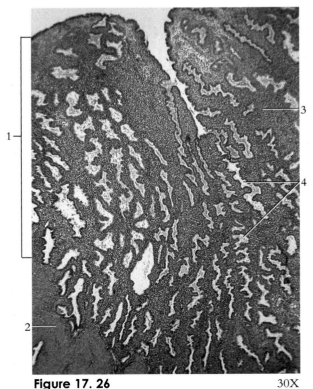

Figure 17. 26   30X

Uterus: secretory phase.
1. Endometrium
2. Myometrium
3. Endometrial stroma
4. Endometrial uterine glands (now saw-toothed)

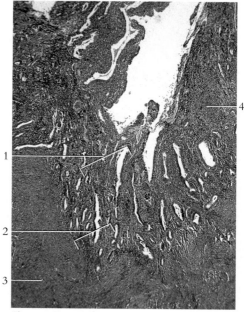

Figure 17. 27   30X

Uterus: menstrual phase.
1. Disintegrating
   lamina functionalis
2. Lamina basalis
   (still intact)
3. Myometrium
4. Blood in stroma

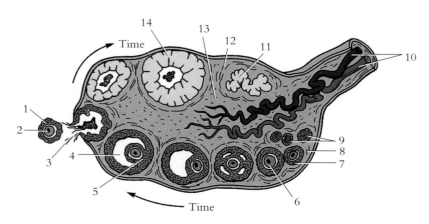

Figure 17.28

Structure of the ovary.
1. Corona radiata
2. Secondary oocyte
3. Ovulation
4. Follicular fluid within antrum
5. Cumulus oophorus
6. Oocyte
7. Follicular cells
8. Germinal epithelium
9. Primary follicles
10. Ovarian vessels
11. Corpus albicans
12. Ovarian cortex
13. Ovarian medulla
14. Corpus luteum

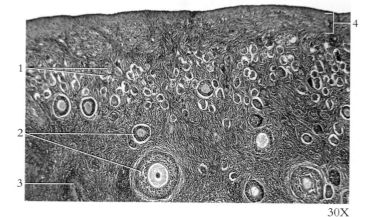

30X

Figure 17.29

Ovary.
1. Primordial follicles
2. Primary follicles
3. Atretic follicle
4. Tunical albuginea

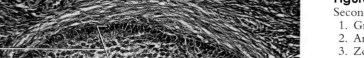

200X

Figure 17.30

Secondary follicle.
1. Granulosa cells
2. Antrum
3. Zona pellucida
4. Oocyte
5. Theca interna
6. Theca externa

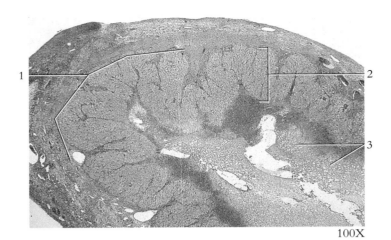

Figure 17. 31
Corpus luteum.
1. Corpus luteum
2. Wall of corpus luteum
3. Former follicular antrum

100X

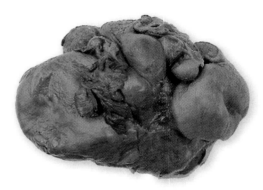

Figure 17. 32
A human ovary showing ovarian cysts.

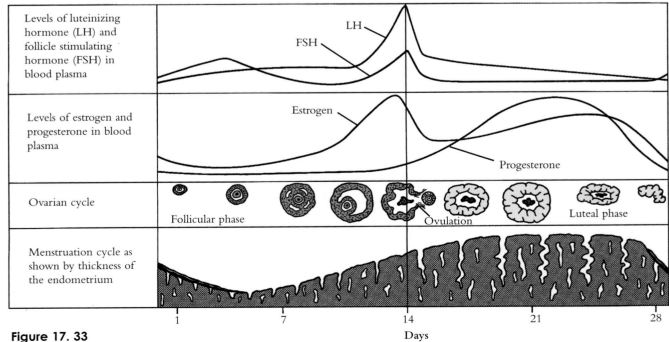

Levels of luteinizing hormone (LH) and follicle stimulating hormone (FSH) in blood plasma	
Levels of estrogen and progesterone in blood plasma	
Ovarian cycle	
Menstruation cycle as shown by thickness of the endometrium	

LH
FSH
Estrogen
Progesterone
Follicular phase
Ovulation
Luteal phase

1   7   14   21   28

Days

Figure 17. 33
Female ovarian and menstruation cycle.

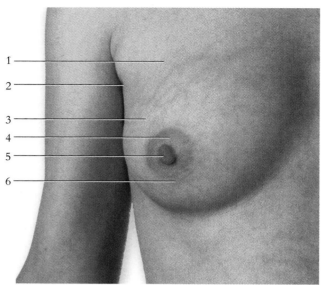

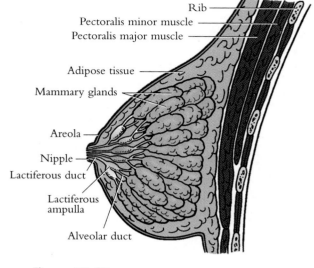

Figure 17. 34
Surface anatomy of the female breast.
1. Pectoralis major muscle
2. Axilla
3. Lateral process of breast
4. Areola
5. Nipple
6. Breast (containing mammary glands)

Figure 17. 35
Mammary gland.

Rib
Pectoralis minor muscle
Pectoralis major muscle
Adipose tissue
Mammary glands
Areola
Nipple
Lactiferous duct
Lactiferous ampulla
Alveolar duct

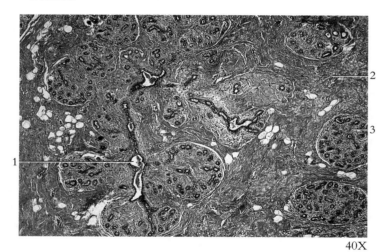

40X

Figure 17.36
Mammary glands, (non–lactating glands).
1. Interlobular duct
2. Interlobular connective tissue
3. Lobule of glandular tissue

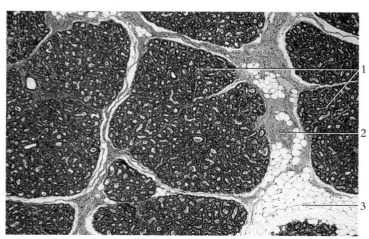

400X

Figure 17. 37
Mammary glands, (lactating glands).
1. Lobules of glandular tissue
2. Intralobular connective tissue
3. Adipose cells

The period of human pregnancy, which generally requires 38 weeks, is known as **gestation**. **Morphogenesis** is the sequence of changes that occur in the formation of the baby's body structures. Although gestation is frequently discussed chronologically as trimesters, prenatal development is more accurately divided morphogenically into three periods based on structural changes. On this basis, the **pre-embryonic period** includes the first two weeks following fertilization, the **embryonic period** includes the following six weeks, and the **fetal period** includes the final 30 weeks.

The events of the two-week pre-embryonic period include transportation of the fertilized egg, or **zygote**, through the uterine tube, mitotic divisions, implantation, and the formation of primordial embryonic tissue (fig. 18.1). Implantation begins between the fifth and seventh day and is made possible by the secretion of enzymes that digest a portion of the endometrium of the uterus. During implantation, the **trophoblast cells** secrete **human chorionic gonadotrophin (hCG)**, which prevents the breakdown of the endometrium and menstruation. The trophoblast cells also participate in the formation of the placenta.

The events of the six-week embryonic period include the differentiation of the germ layers into specific body organs and the formation of the extraembryonic membranes, including the **placenta, umbilical cord, amnion, yolk sac, allantois**, and **chorion**.

A small amount of tissue differentiation and organ development occurs during the fetal period, but for the most part fetal development is primarily limited to body growth. Labor and parturition (childbirth) are the culmination of gestation and require the action of **oxytocin** from the mother's pituitary gland, and **prostaglandins**, produced in her uterus.

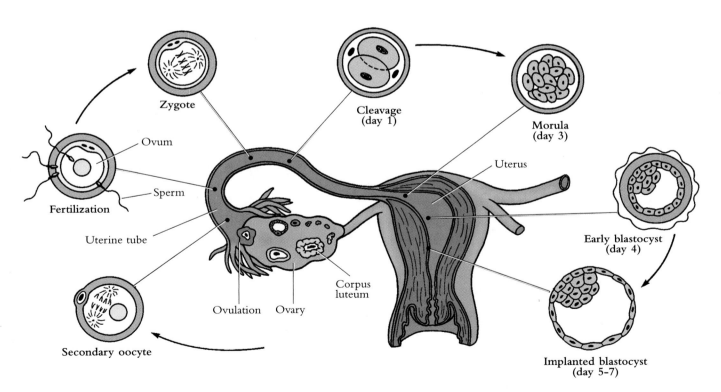

Figure 18.1
Events of ovulation, fertilization, and implantation.

Figure 18.2
A 8.1 week intrauterine pregnancy. Marks are indicated on the image to denote the crown and rump. The crown–rump length of this embryo is 18 mm. Ultrasonography, produced by a mechanical vibration of high frequency, produces a safe, high resolution of fetal structure. Most ultrasound scans are obtained on fetuses older than 12 weeks.

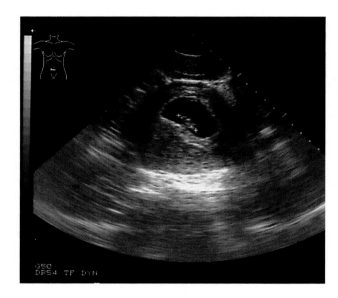

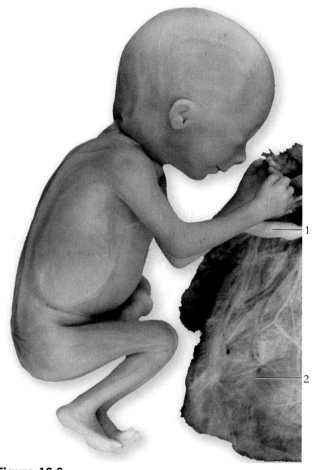

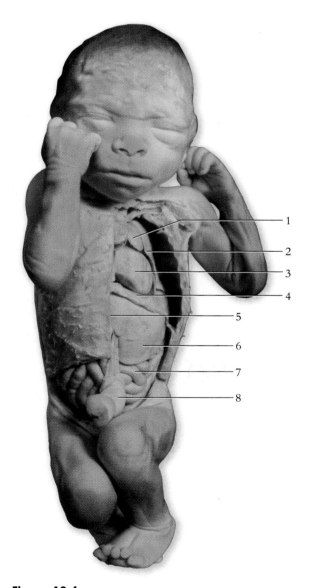

Figure 18.3
A human fetus at aproximately 24 weeks.
1. Umbilical cord
2. Placenta

Figure 18.4
A human fetus at aproximately 28 weeks.
1. Thymus   5. Falciform ligament
2. Lung   6. Liver
3. Heart   7. Small intestine
4. Diaphragm   8. Umbilical cord

Figure 18.5

Thoracic and abdominal viscera of a human fetus at approximately 28 weeks.

1. Pectoralis major m.
2. Pericardium (cut)
3. Pericardial cavity
4. Falciform ligament
5. External abdominal oblique m.
6. Umbilicus
7. Umbilical vein
8. Thymus
9. Lung
10. Pleural cavity
11. Heart
12. Thoracic wall
13. Diaphragm
14. Liver
15. Peritoneal cavity
16. Abdominal wall
17. Small intestine
18. Umbilical cord

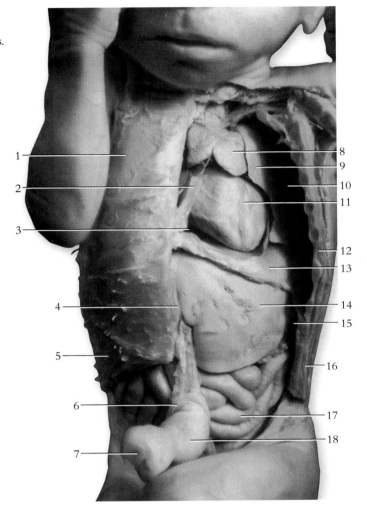

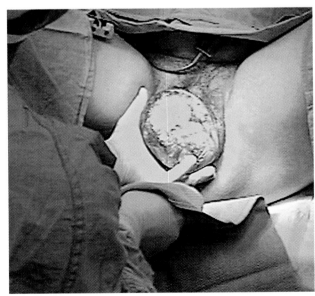

Figure 18.6
Parturition, or childbirth.

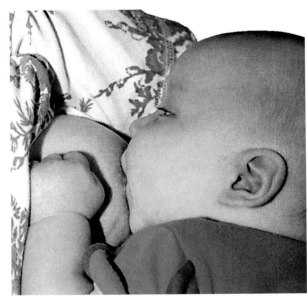

Figure 18.7
Nursing.

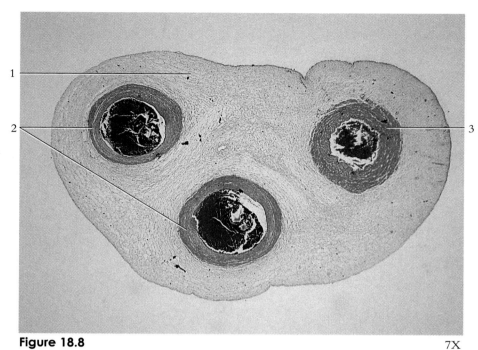

Figure 18.8   7X
Umbilical cord.

1. Mucoid connective tissue   2. Umbilical arteries
   (Wharton's jelly)   3. Umbilical vein

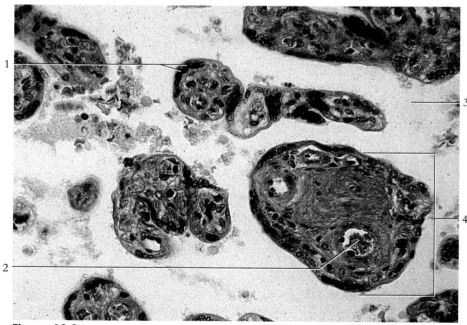

Figure 18.9   300X
Placenta.

1. Nuclei of syncytiotrophoblast   3. Intervillus space containing
2. Blood vessel containing fetal   maternal blood cells
   blood cells   4. Choronic villus

Embalmed cats purchased from biological supply houses are excellent specimens for dissecting and learning basic mammalian anatomy. Before the muscles and viscera of a cat can be studied, the specimen's skin has to be removed according to the following suggested guidelines.

1. Place the cat on a dissecting tray dorsal side up. Using a sharp scalpel, make a short, shallow incision through the skin across the nape of the neck. With your scissors, continue a dorsal midline incision forward over the skull and down the back to about two inches onto the tail. Sever the tail with bone shears or a saw and discard.

2. Make a shallow incision around the neck and down each foreleg to the paws. Continue a circular incision around each wrist. Beginning at the base of the tail, make incisions down each of the hind legs to the ankles. Make a circular cut around each ankle.

3. Carefully remove the skin, using your fingers or a blunt probe to separate the skin from underlying connective tissue. Where it is necessary to use a scalpel, keep the cutting edge directed toward the skin away from the muscle. If your specimen is a male, make an incision around the genitalia, leaving the skin intact. If your specimen is a female, the mammary glands will appear as longitudinal, glandular masses along the ventral side of the abdomen and thorax. They should be removed with the skin.

4. After the specimen is skinned, remove the excess fat and connective tissue to expose the underlying muscles. Make certain that the muscles are separated along their natural boundaries. If a transection of a muscle is necessary, isolate the muscle from its attached connective tissue and make a clean cut across the belly of the muscle, leaving the origin and insertion intact.

5. At the end of the laboratory period, wrap your specimen in muslin cloth and store it in a tight, heavy-duty plastic bag. Wet your specimen from time to time with a preservative solution (usually 2%–3% phenol). Caution needs to be taken when using a phenol wetting solution as it is caustic and poisonous if misused or used in a concentrated form.

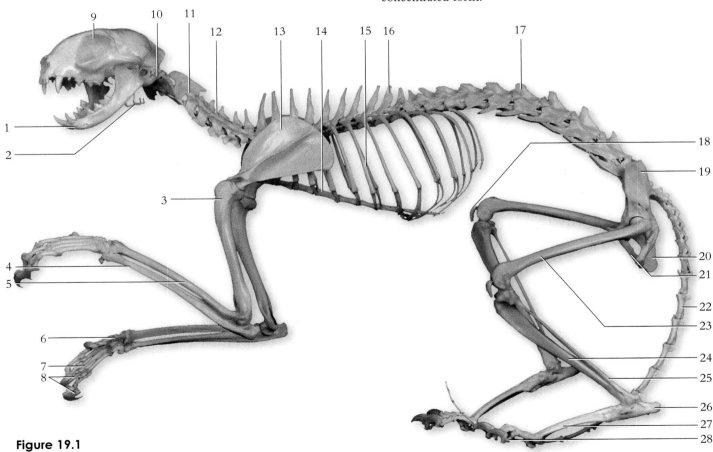

Figure 19.1
Cat skeleton.

1. Mandible	7. Metacarpal bones	13. Scapula	19. Ilium	25. Fibula
2. Hyoid bone	8. Phalanges	14. Sternum	20. Ischium	26. Tarsal bones
3. Humerus	9. Skull	15. Rib	21. Pubis	27. Metatarsal bones
4. Ulna	10. Atlas	16. Thoracic vertebra	22. Caudal vertebra	28. Phalanges
5. Radius	11. Axis	17. Lumbar vertebra	23. Femur	
6. Carpal bones	12. Cervical vertebra	18. Patella	24. Tibia	

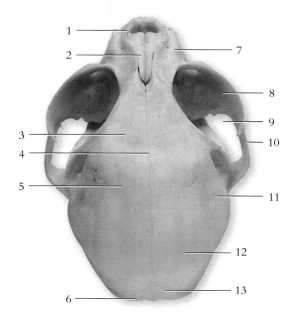

Figure 19.2

A dorsal view of a cat skull.

1. Premaxilla
2. Nasal bone
3. Frontal bone
4. Sagittal suture
5. Coronal suture
6. Nuchal crest
7. Maxilla
8. Zygomatic (malar) bone
9. Orbit
10. Zygomatic arch
11. Temporal bone
12. Parietal bone
13. Interparietal bone

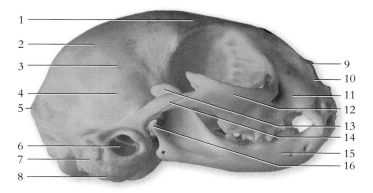

Figure 19.3

A lateral view of a cat skull.

1. Frontal bone
2. Parietal bone
3. Squamosal suture
4. Temporal bone
5. Nuchal crest
6. External acoustic meatus
7. Mastoid process
8. Tympanic bulla
9. Nasal bone
10. Premaxilla bone
11. Maxilla
12. Zygomatic (malar) bone
13. Coronoid process of mandible
14. Zygomatic arch
15. Mandible
16. Condylar process of mandible

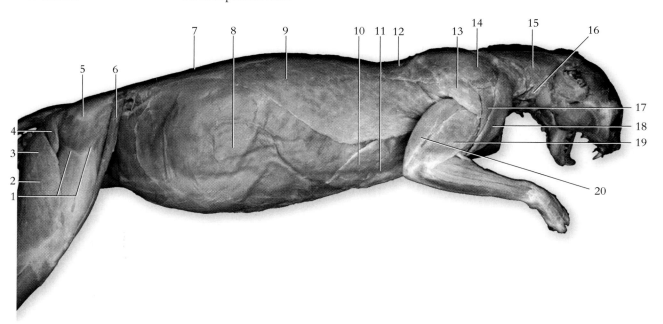

Figure 19.4

A lateral view of the superficial muscles of the cat.

1. Tensor fasciae latae m.
2. Biceps femoris m.
3. Caudofemoralis m.
4. Gluteus maximus m.
5. Gluteus medius m.
6. Sartorius m.
7. Lumbodorsal fascia
8. External abdominal oblique m.
9. Latissimus dorsi m.
10. Xiphihumeralis m.
11. Pectoralis minor m.
12. Spinotrapezius m.
13. Spinodeltoid m.
14. Acromiotrapezius m.
15. Clavotrapezius m.
16. Sternomastoid m.
17. Acromiodeltoid m.
18. Clavobrachialis m.
19. Lateral head of triceps brachii m.
20. Long head of triceps brachii m.

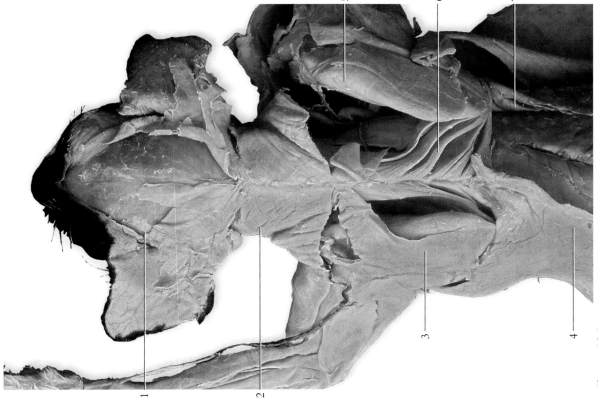

Figure 19.6
A dorsal view of the cat neck and thorax.
1. Temporalis m.
2. Clavotrapezius m.
3. Acromiotrapezius m.
4. Latissimus dorsi m.
5. Supraspinatus m.
6. Rhomboideus m.
7. Serratus anterior m.

Figure 19.5
A dorsal view of the superficial muscles of the cat.
1. Lateral head of triceps brachii m.
2. Acromiotrapezius m.
3. Latissimus dorsi m.
4. Lumbodorsal fascia
5. Sacrospinalis m.
6. Gluteus medius m.
7. Caudal m..
8. Supraspinatus m
9. Rhomboideus m.
10. Serratus anterior m.
11. Latissimus dorsi m.
12. Gluteus maximus m.

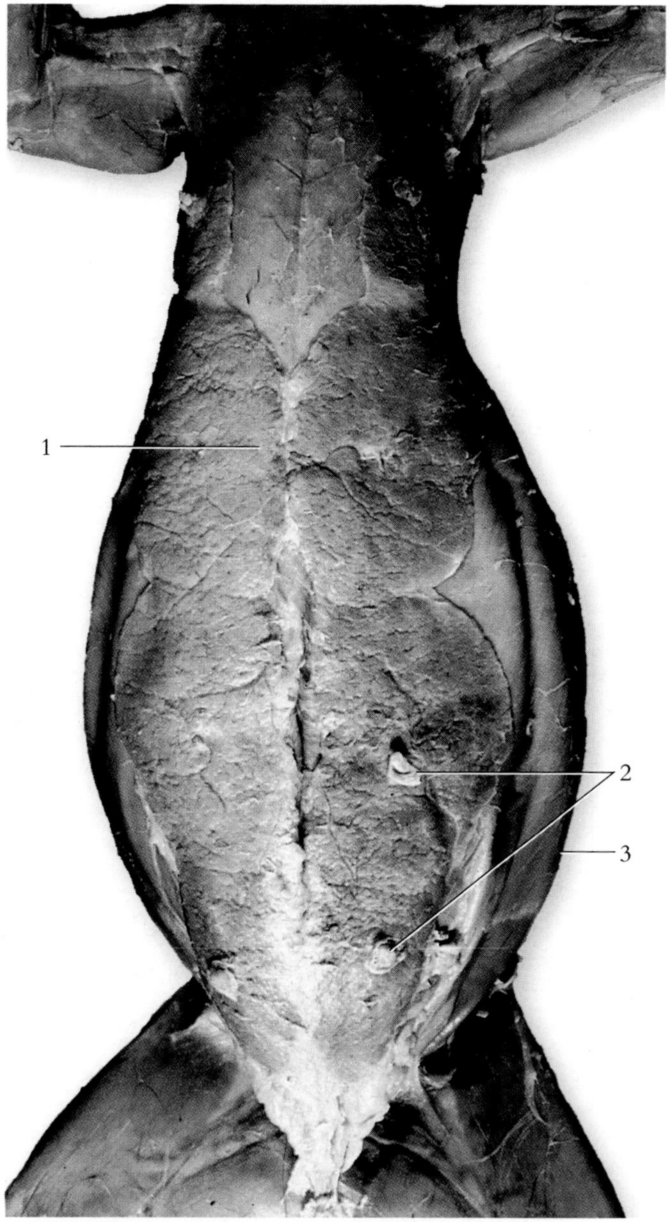

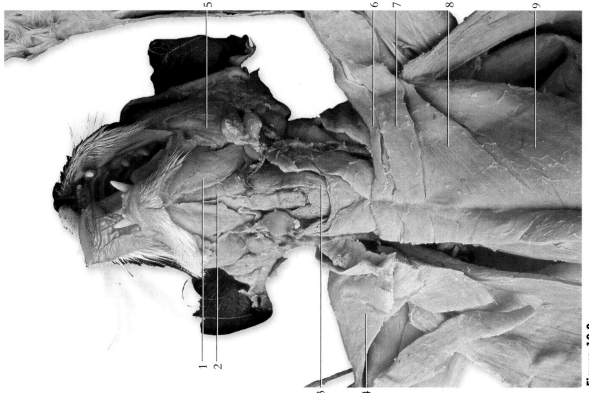

Figure 19.8

A ventral view of the neck and thorax.

1. Digastric m.
2. Mylohyoid m.
3. Sternomastoid m.
4. Clavotrapezius m.
5. Masseter m.

6. Clavobrachialis m.
7. Pectoantebrachialis m.
8. Pectoralis major m.
9. Pectoralis minor m.

Figure 19.7

A superficial view of the ventral trunk.

1. Mammary glands
2. Nipples

3. External abdominal oblique m.

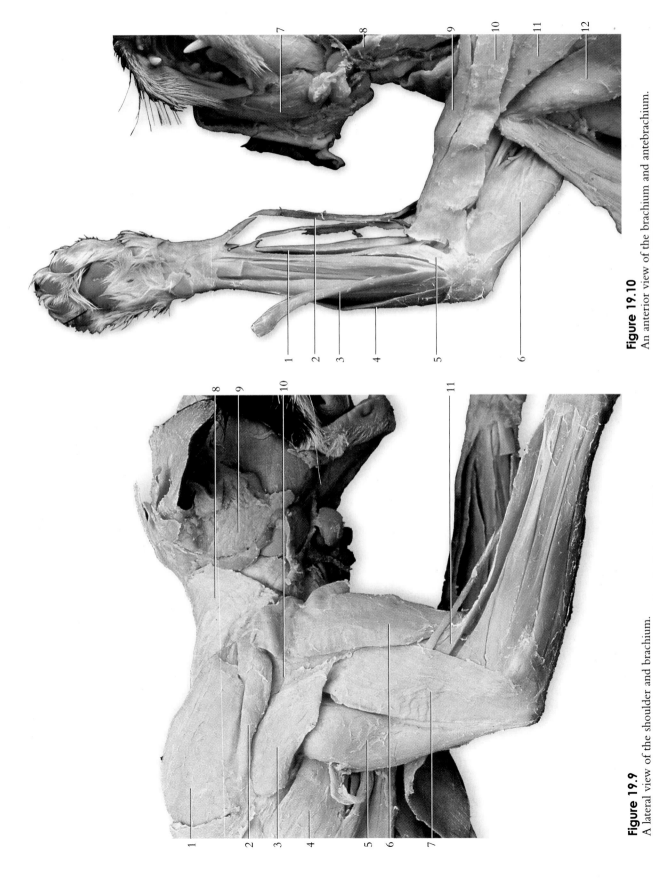

Figure 19.10

An anterior view of the brachium and antebrachium.

1. Extensor carpi radialis longus m.
2. Brachioradialis m.
3. Palmaris longus m. (cut)
4. Flexor carpi ulnaris m.
5. Pronator teres m.
6. Epitrochlearis
7. Masseter m.
8. Sternomastoid m.
9. Clavobrachialis m.
10. Pectoantebrachialis m.
11. Pectoralis major m.
12. Pectoralis minor m.

Figure 19.9

A lateral view of the shoulder and brachium.

1. Acromiotrapezius m.
2. Levator scapulae ventralis m.
3. Spinodeltoid m.
4. Latissimus dorsi m.
5. Long head of triceps brachii m.
6. Clavobrachialis m.
7. Lateral head of triceps brachii m.
8. Clavotrapezius m.
9. Parotid gland
10. Acromiodeltoid m.
11. Brachioradialis m.

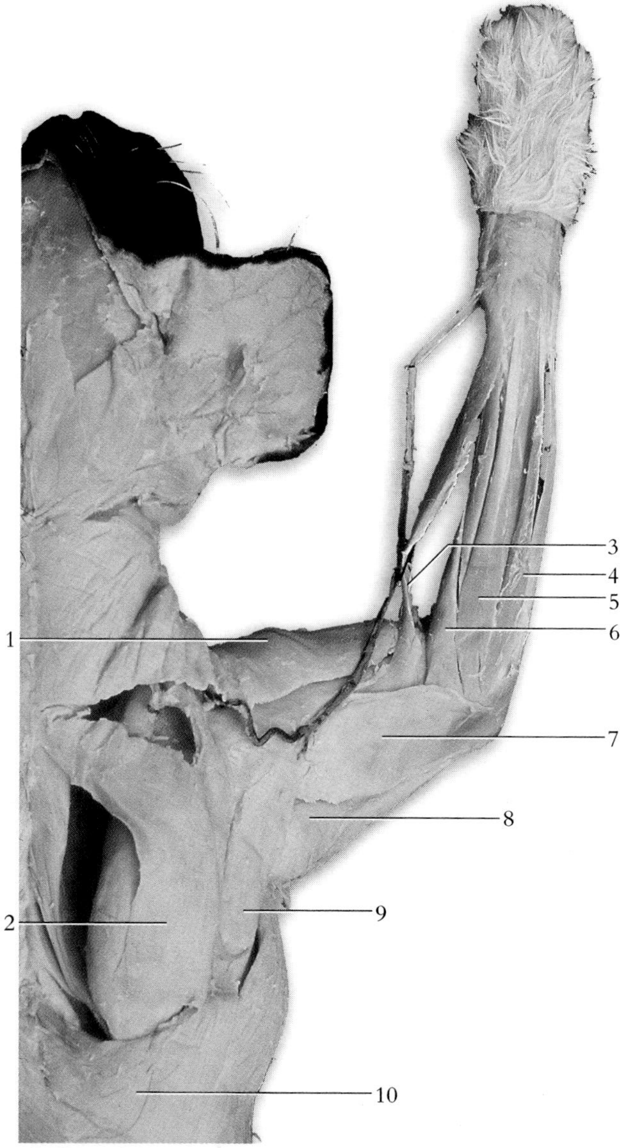

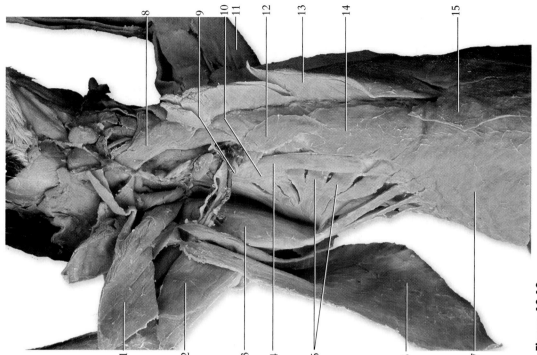

Figure 19.12

An anterior view of the trunk.

1. Pectoralis minor (cut)
2. Epitrochlearis m.
3. Subscapularis m.
4. Scalenus medius m.
5. Serratus anterior m.
6. Latissimus dorsi m. (cut)
7. External abdominal oblique m.
8. Sternomastoid m.
9. Scalenus anterior m.
10. Scalenus posterior m.
11. Epitrochlearis m.
12. Transverse costarum m.
13. Pectoralis minor m. (cut)
14. Rectus abdominis m.
15. Xiphihumeralis m. (cut)

Figure 19.11

A posterior view of the brachium and antebrachium.

1. Clavobrachialis m.
2. Acromiotrapezius m.
3. Brachioradialis m.
4. Extensor digitorum lateralis m.
5. Extensor digitorum communis m.
6. Extensor carpi radialis longus m.
7. Lateral head of triceps brachii m.
8. Long head of triceps brachii m.
9. Spinodeltoid m.
10. Latissimus dorsi m.

Figure 19.13
A lateral view of the trunk.
1. Internal abdominal oblique m.
2. Tensor fascia latae
3. Caudofemoris m.
4. Vastus lateralis m.
5. Sartorius m.
6. External abdominal oblique m.

7. Latissimus dorsi m.
8. Spinodeltoid m.
9. Transverse abdominis m.
10. Serratus anterior m.
11. Long head of triceps brachii m.

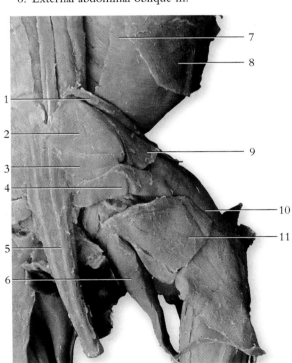

Figure 19.14
A lateral view of the superficial thigh.
1. Sartorius m.
2. Gluteus medius m.
3. Gluteus maximus m.
4. Caudofemoris m.
5. Caudal m.
6. Semitendinosus m.
7. Internal abdominal oblique m.
8. External abdominal oblique m.
9. Tensor fascia latae (cut)
10. Vastus lateralis m.
11. Biceps femoris m.

Figure 19.15
A medial view of the thigh and leg.
1. Sartorius m.
2. Tensor fascia latae m.
3. Vastus lateralis m.
4. Rectus femoris m.
5. Vastus medialis m.
6. Flexor digitorum longus m.
7. Tibialis anterior m.
8. Rectus abdominus m.
9. Adductor longus m.
10. Adductor femoris m.
11. Semimembranosus m.
12. Gracilis m. (cut)
13. Tendo calcaneus (Achilles tendon)

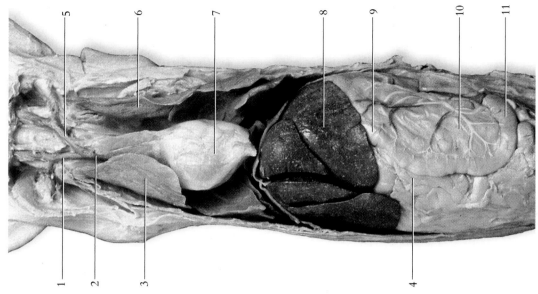

Figure 19.17
Intact viscera.

1. Right brachiocephalic vein
2. Superior vena cava
3. Right lung
4. Greater omentum
5. Left brachiocephalic vein
6. Left lung
7. Heart
8. Liver
9. Stomach
10. Mesentery
11. Small intestine

Figure 19.16
A lateral view of the thigh and leg.

1. Gluteus medius m.
2. Gluteus maximus m.
3. Caudofemoralis m.
4. Sciatic nerve
5. Semimembranosus m.
6. Semitendinosus m.
7. Gastrocnemius m.
8. Tendo calcaneus
9. Vastus lateralis m.
10. Adductor femoris m.
11. Tenuissimus m.
12. Biceps femoris m. (cut)
13. Soleus m.
14. Peroneal m.

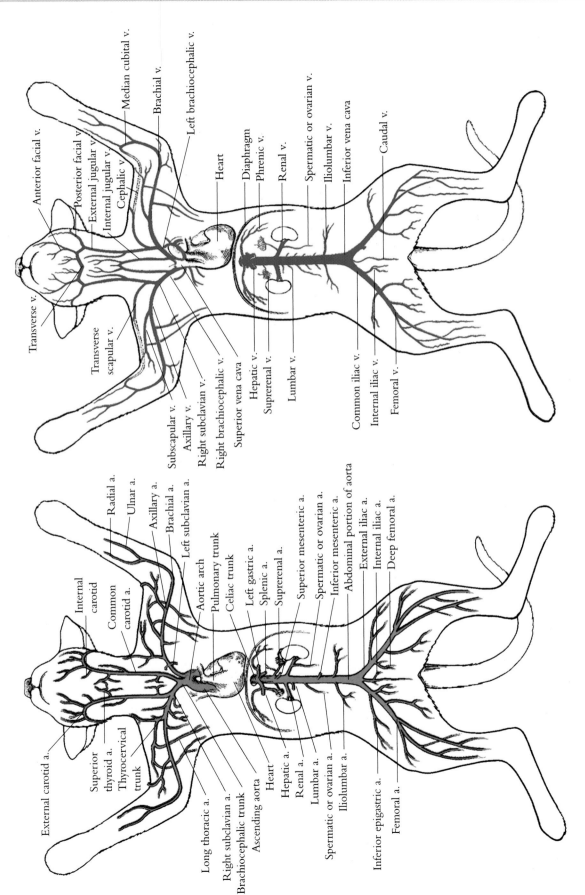

Figure 19.19
Principal veins of the cat, ventral view. (v=vein)

Anterior facial v.
Posterior facial v.
External jugular v.
Internal jugular v.
Cephalic v.
Transverse v.
Transverse scapular v.
Median cubital v.
Brachial v.
Left brachiocephalic v.
Heart
Diaphragm
Phrenic v.
Renal v.
Spermatic or ovarian v.
Iliolumbar v.
Inferior vena cava
Caudal v.
Subscapular v.
Axillary v.
Right subclavian v.
Right brachiocephalic v.
Superior vena cava
Hepatic v.
Suprerenal v.
Lumbar v.
Common iliac v.
Internal iliac v.
Femoral v.

Figure 19.18
Principal arteries of the cat, ventral view. (a=artery)

Internal carotid
Common carotid a.
Radial a.
Ulnar a.
Axillary a.
Brachial a.
Left subclavian a.
Aortic arch
Pulmonary trunk
Celiac trunk
Left gastric a.
Splenic a.
Suprerenal a.
Superior mesenteric a.
Spermatic or ovarian a.
Inferior mesenteric a.
Abdominal portion of aorta
External iliac a.
Internal iliac a.
Deep femoral a.
External carotid a.
Superior thyroid a.
Thyrocervical trunk
Long thoracic a.
Right subclavian a.
Brachiocephalic trunk
Ascending aorta
Heart
Hepatic a.
Renal a.
Lumbar a.
Spermatic or ovarian a.
Iliolumbar a.
Inferior epigastric a.
Femoral a.

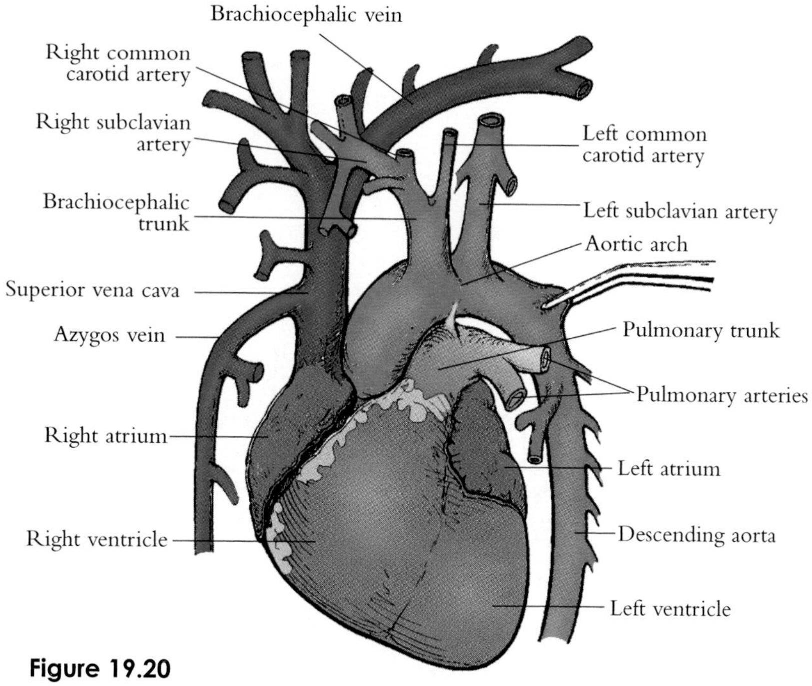

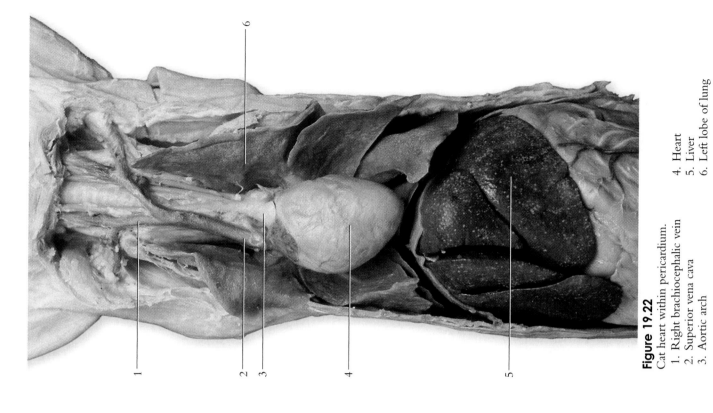

Figure 19.22
Cat heart within pericardium.
1. Right brachiocephalic vein
2. Superior vena cava
3. Aortic arch
4. Heart
5. Liver
6. Left lobe of lung

Brachiocephalic vein

Right common carotid artery
Left common carotid artery
Right subclavian artery
Left subclavian artery
Aortic arch
Brachiocephalic trunk
Pulmonary trunk
Superior vena cava
Pulmonary arteries
Azygos vein
Left atrium
Right atrium
Descending aorta
Right ventricle
Left ventricle

Figure 19.20
The cat heart and associated vessels.

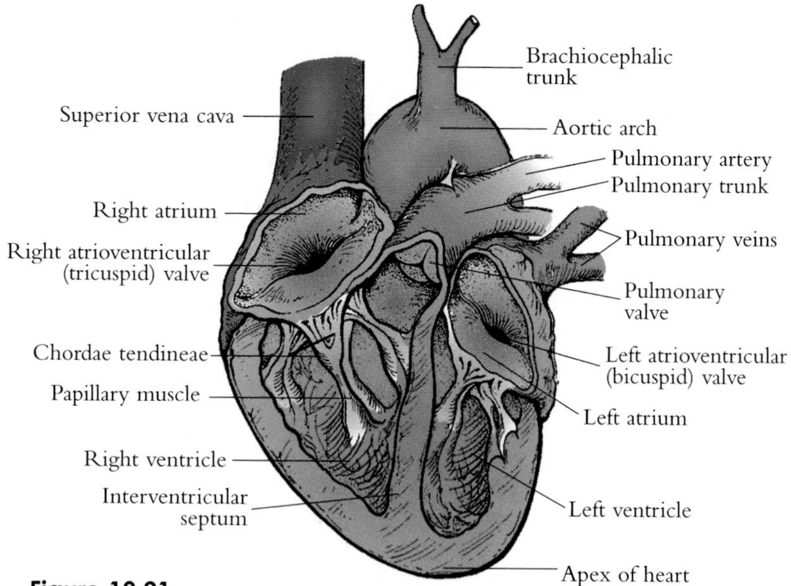

Brachiocephalic trunk
Aortic arch
Pulmonary artery
Pulmonary trunk
Pulmonary veins
Pulmonary valve
Left atrioventricular (bicuspid) valve
Left atrium
Superior vena cava
Left ventricle
Right atrium
Apex of heart
Right atrioventricular (tricuspid) valve
Chordae tendineae
Papillary muscle
Right ventricle
Interventricular septum

Figure 19.21
Internal anatomy of the cat heart.

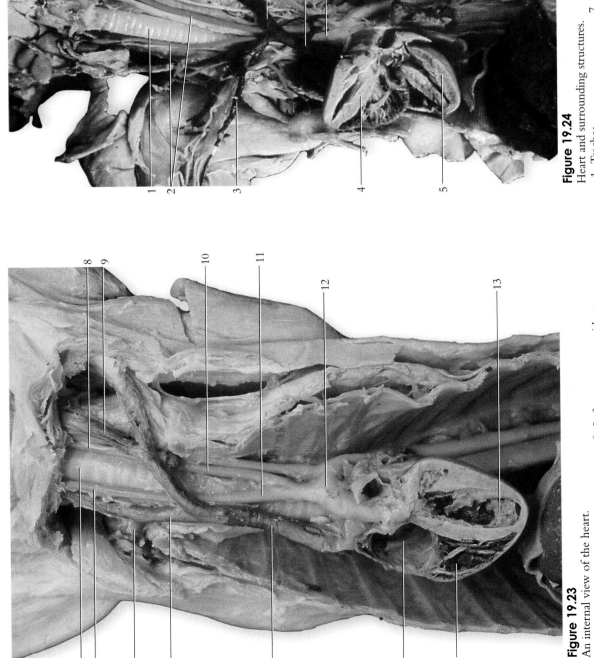

Figure 19.24
Heart and surrounding structures.

1. Trachea
2. Common carotid arteries
3. Axillary vein
4. Heart (cut)
5. Left ventricle
6. Vagus nerve
7. External jugular vein
8. Left subclavian vein
9. Cranial (superior) vena cava
10. Brachiocephalic trunk
11. Thoracic aorta

Figure 19.23
An internal view of the heart.

1. Trachea
2. Right common carotid artery
3. Right subclavian vein
4. Right brachiocephalic vein
5. Cranial (superior) vena cava
6. Right atrium
7. Right ventricle
8. Left common carotid artery
9. Internal jugular vein
10. Brachiocephalic artery
11. Right common carotid artery
12. Aortic arch
13. Left ventricle

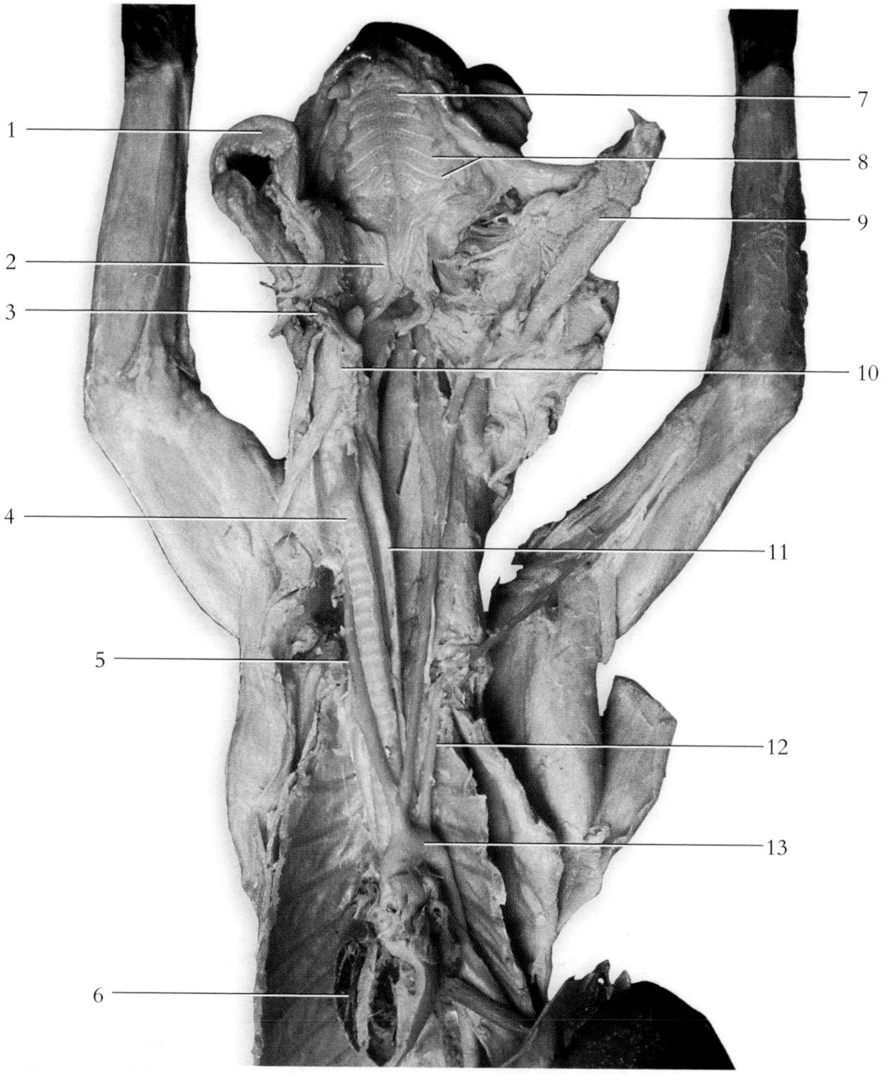

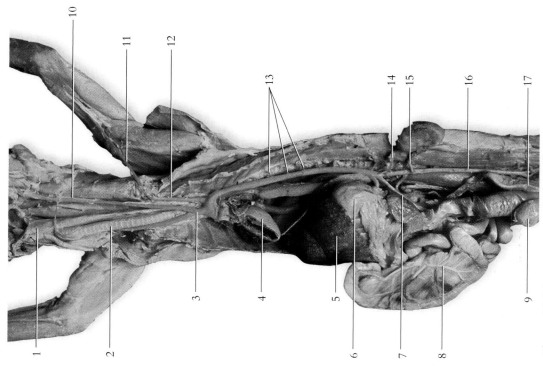

Figure 19.25

Upper gastrointestinal and respiratory structures of the a cat.

1. Tongue
2. Soft palate
3. Epiglottis
4. Trachea
5. Right common carotid artery
6. Heart (cut)
7. Hard palate
8. Palatal rugae
9. Mandible (cut)
10. Larynx
11. Esophagus
12. Left subclavian artery
13. Aortic arch

Figure 19.26

An anterior view of the arteries and veins of the trunk of a cat.

1. Larynx
2. Trachea
3. Brachiocephalic trunk
4. Heart (cut)
5. Liver
6. Stomach
7. Superior mesenteric artery
8. Superior mesenteric vein
9. Urinary bladder
10. Left common carotid artery
11. Left axillary artery
12. Left subclavian artery
13. Intercostal arteries
14. Suprarenal artery
15. Renal artery
16. Abdominal aorta
17. Left horn of uterus

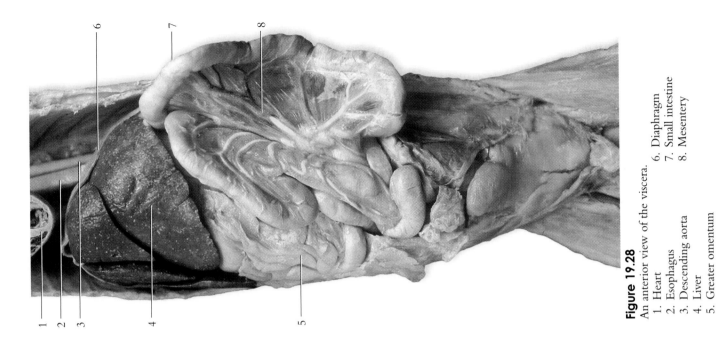

Figure 19.28
An anterior view of the viscera.
1. Heart
2. Esophagus
3. Descending aorta
4. Liver
5. Greater omentum
6. Diaphragm
7. Small intestine
8. Mesentery

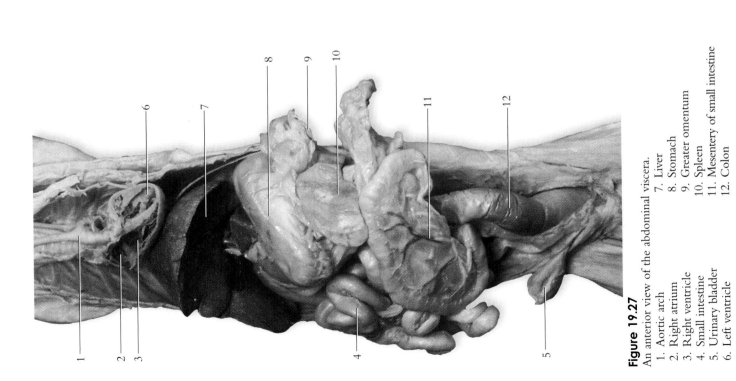

Figure 19.27
An anterior view of the abdominal viscera.
1. Aortic arch
2. Right atrium
3. Right ventricle
4. Small intestine
5. Urinary bladder
6. Left ventricle
7. Liver
8. Stomach
9. Greater omentum
10. Spleen
11. Mesentery of small intestine
12. Colon

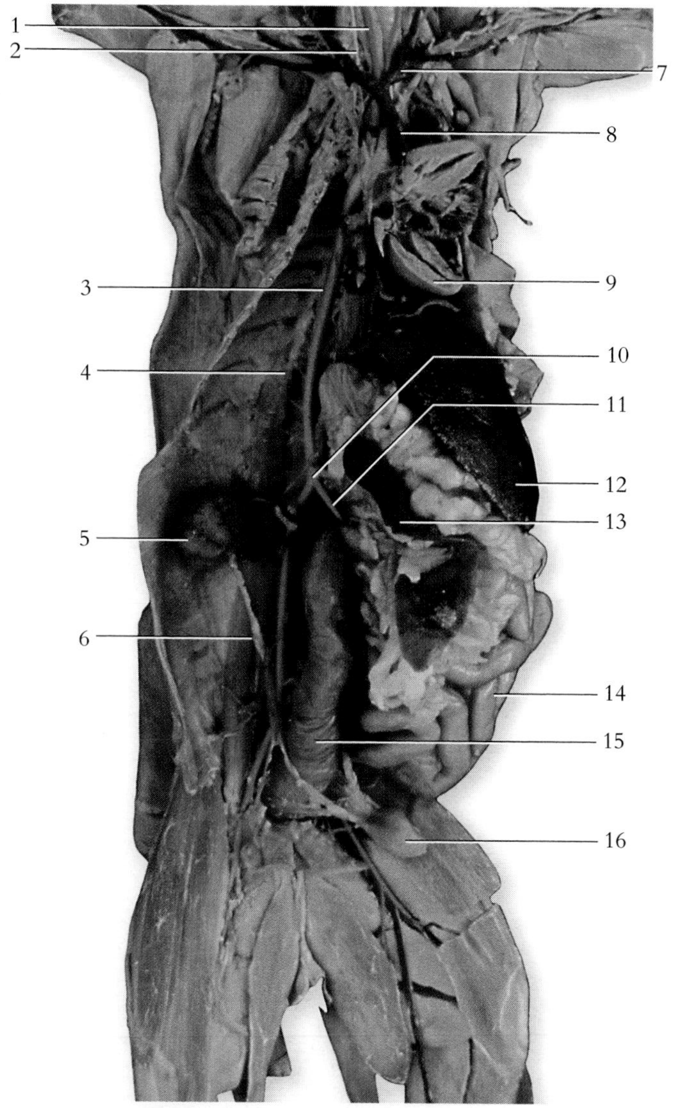

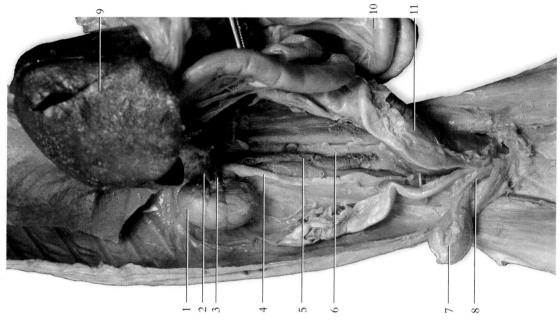

Figure 19.30
The urinary system.
1. Kidney
2. Renal artery
3. Renal vein
4. Ureter
5. Caudal (inferior) vena cava
6. Abdominal aorta
7. Urinary bladder
8. Urethra
9. Liver
10. Small intestine
11. Colon

Figure 19.29
An anterior view of the deep structures of the trunk.
1. Right common carotid artery
2. Vagus nerve
3. Thoracic aorta
4. Intercostal artery
5. Kidney
6. Ureter
7. Left brachiocephalic vein
8. Cranial (superior) vena cava
9. Heart
10. Celiac trunk
11. Superior mesenteric artery
12. Liver
13. Spleen
14. Small intestine
15. Colon
16. Urinary bladder

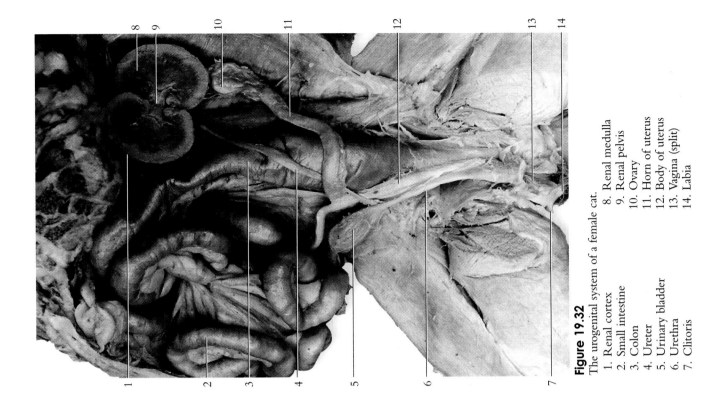

Figure 19.32

The urogenital system of a female cat.

1. Renal cortex
2. Small intestine
3. Colon
4. Ureter
5. Urinary bladder
6. Urethra
7. Clitoris
8. Renal medulla
9. Renal pelvis
10. Ovary
11. Horn of uterus
12. Body of uterus
13. Vagina (split)
14. Labia

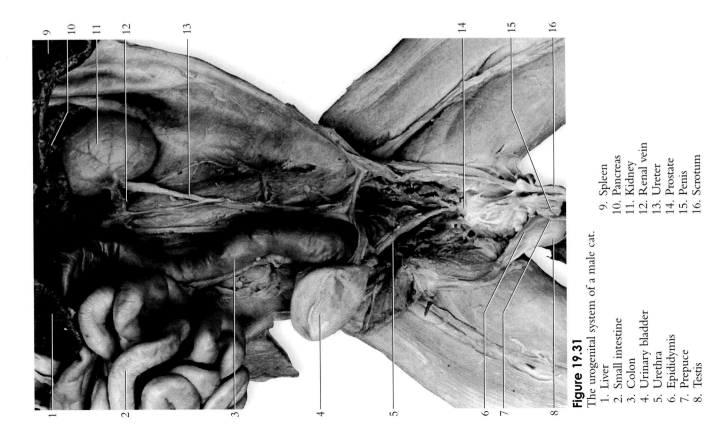

Figure 19.31

The urogenital system of a male cat.

1. Liver
2. Small intestine
3. Colon
4. Urinary bladder
5. Urethra
6. Epididymis
7. Prepuce
8. Testis
9. Spleen
10. Pancreas
11. Kidney
12. Renal vein
13. Ureter
14. Prostate
15. Penis
16. Scrotum

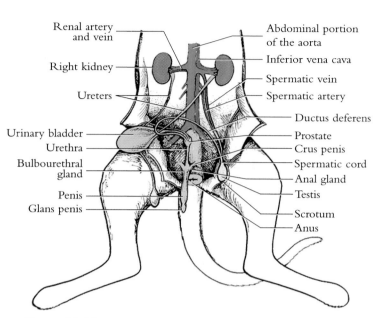

Renal artery and vein

Right kidney

Ureters

Urinary bladder

Urethra

Bulbourethral gland

Penis

Glans penis

Abdominal portion of the aorta

Inferior vena cava

Spermatic vein

Spermatic artery

Ductus deferens

Prostate

Crus penis

Spermatic cord

Anal gland

Testis

Scrotum

Anus

Figure 19.33
Male urinary and reproductive systems of the cat.

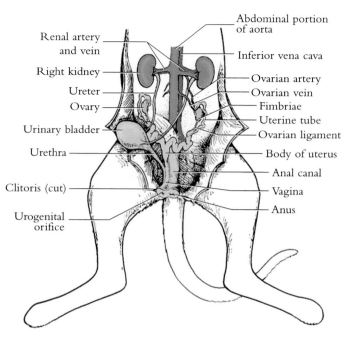

Renal artery and vein

Right kidney

Ureter

Ovary

Urinary bladder

Urethra

Clitoris (cut)

Urogenital orifice

Abdominal portion of aorta

Inferior vena cava

Ovarian artery

Ovarian vein

Fimbriae

Uterine tube

Ovarian ligament

Body of uterus

Anal canal

Vagina

Anus

Figure 19.34
Female urinary and reproductive systems of the cat.

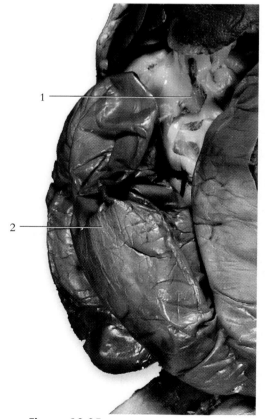

1

2

Figure 19.35
Abdominal cavity of a pregnant cat.
  1. Greater omentum
  2. Right horn of uterus

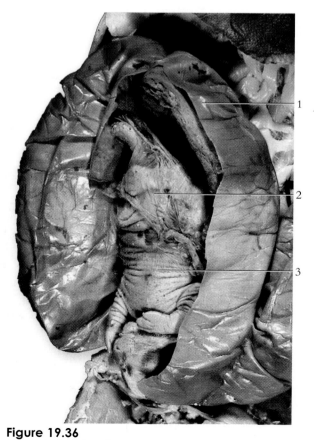

1

2

3

Figure 19.36
Abdominal cavity of a pregnant cat.
  1. Uterine wall (cut)
  2. Amniotic sac enclosing fetus
  3. Fetus

Because much can be learned from dissecting embalmed fetal pig specimens, they are frequently utilized in anatomy laboratories. Fetal pigs are purchased from biological supply houses and are specially prepared for dissection. Excess embalming fluid should be drained from the packaged specimen prior to dissection.

Examine your specimen and identify the **umbilical cord** attached to the ventral surface of the abdomen. Locate the two rows of **teats** that extend the length of the abdomen. Determine the sex of your specimen. A male has a **scrotal sac** in the pelvic region of the body between the hind legs and a **urogenital opening** just caudal to the umbilical cord. The **penis** can be palpated as a muscular tubular structure just underneath the skin along the midline proceeding caudally from the urogenital opening. A female has a small fleshy **genital papilla** projecting from the urogenital opening, which is located immediately ventral to the **anal opening**.

Before the muscles and viscera of a fetal pig can be studied, the specimen's skin has to be removed according to the following suggested guidelines.

1. Place your specimen on a dissecting tray ventral side up. Using a sharp scalpel, make a shallow incision through the skin extending from the chin caudally to the umbilical cord. Carefully continue your cut around one side of the umbilical cord. If your specimen is a male, make a diagonal cut from the umbilical cord to the scrotum. If a female, continue a midventral incision from the umbilical cord to the genital papilla. Make an incision around the genitalia and tail.

2. From the midventral incision, extend an incision down the medial surfaces of the forelegs to the hoofs and then do the same for the skin of the hindlegs. Make circular incisions around each of the hoofs. Following the ventral borders of the lower jaws, make extended cuts from the chin dorsolaterally to just below the ears.

3. Grasp the cut edge of the skin and carefully remove it from your specimen. If the skin is difficult to remove, grasp the cut edge of the skin with one hand and push on the muscle with the thumb of the other hand.

4. After the specimen is skinned, the muscles can be seen more easily if the moisture on them is sponged away with a paper towel. The muscles of a fetal pig are extremely delicate and as you proceed to dissect your specimen, make certain that you separate the muscles along their natural boundaries. When transection of a muscle is necessary, carefully isolate the muscle from its attached connective tissue and make a clean cut across the belly of the muscle, leaving the origin and insertion intact.

5. At the end of the laboratory period, wrap your specimen in muslin cloth and store it in a tight, heavy-duty plastic bag. Discard the skin that was removed from your specimen, and the plastic shipment bag. Wet your specimen from time to time with a preservative solution (usually 2-3% phenol). Caution is necessary when using a phenol wetting solution as it is caustic and poisonous if misused or used in a concentrated form.

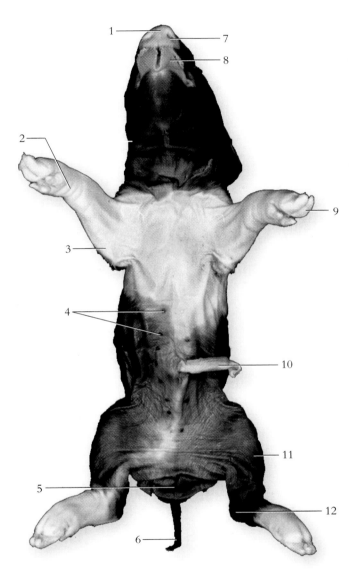

Figure 20.1
A ventral view of the surface anatomy of the fetal pig.

1. Nose	5. Scrotum	9. Hoof of digit
2. Wrist	6. Tail	10. Umbilical cord
3. Elbow	7. Nostril	11. Knee
4. Teats	8. Tongue	12. Ankle

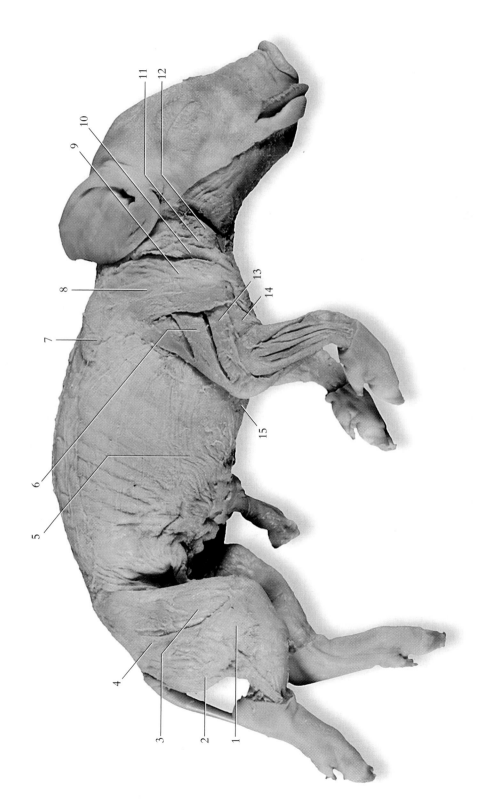

Figure 20.2

Lateral view of superficial musculature of the fetal pig.

1. Biceps femoris m.
2. Semitendinosus m.
3. Tensor fasciae latae m.
4. Gluteus medius m.
5. External abdominal oblique m.
6. Triceps brachii m. (long head)
7. Trapezius m.
8. Deltoid m.
9. Supraspinatus m.
10. Cleidooccipitalis m.
11. Cleidomastoid m.
12. Sternocephalicus m.
13. Triceps brachii m. (lateral head)
14. Brachialis m.
15. Pectoralis profundus m.

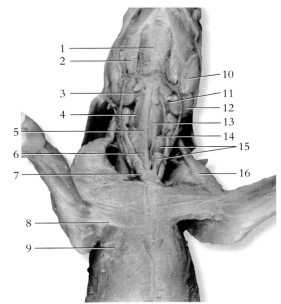

Figure 20.3
Ventral view of superficial muscles of neck and upper torso.
1. Mylohyoid m.
2. Digastric m.
3. Stylohyoid m.
4. Omohyoid m.
5. Sternohyoid m.
6. Thymus
7. Sternomastoid m.
8. Pectoralis superficialis m.
9. Pectoralis profundus m.
10. Masseter m.
11. Thyrohyoid m.
12. Mandibular gland
13. Larynx
14. Sternothyroid m.
15. Mandibular lymph nodes
16. Brachiocephalic m.

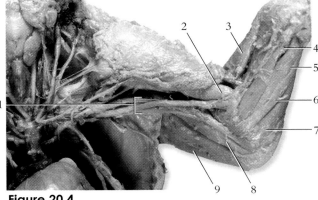

Figure 20.4
Superficial medial muscles of the forelimb.
1. Axillary artery and vein, brachial plexus
2. Biceps brachii m.
3. Extensor carpi radialis m.
4. Flexor carpi radialis m.
5. Flexor digitorum profundus m.
6. Flexor digitorum superficialis m.
7. Flexor carpi ulnaris m.
8. Triceps brachii m. (lateral head)
9. Triceps brachii m. (long head)

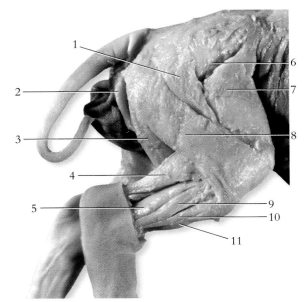

Figure 20.5
A lateral view of the superficial thigh and leg.
1. Gluteus superficialis m.
2. Semitendinosus m.
3. Semimembranosus m.
4. Gastrocnemius m.
5. Extensor digitorum quarti and quinti mm.
6. Gluteus medius m.
7. Tensor fasciae latae m.
8. Biceps femoris m.
9. Peroneus longus m.
10. Peroneus tertius m.
11. Tibialis anterior m.

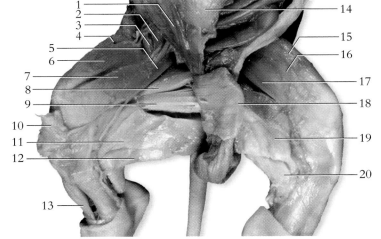

Figure 20.6
Medial muscles of thigh and leg.
1. External abdominal oblique m.
2. Psoas major m.
3. Iliacus m.
4. Tensor fasciae latae m.
5. Sartorius m.
6. Rectus femoris m.
7. Vastus medialis m.
8. Pectineus m.
9. Adductor m.
10. Aponeurosis of gracilis (cut)
11. Semimembranosus m.
12. Semitendinosus m.
13. Tibialis anterior m.
14. Linea alba
15. Rectus femoris m.
16. Vastus medialis m.
17. Sartorius m.
18. Gracilis m. (cut)
19. Gracilis m.
20. Semitendinosus m.

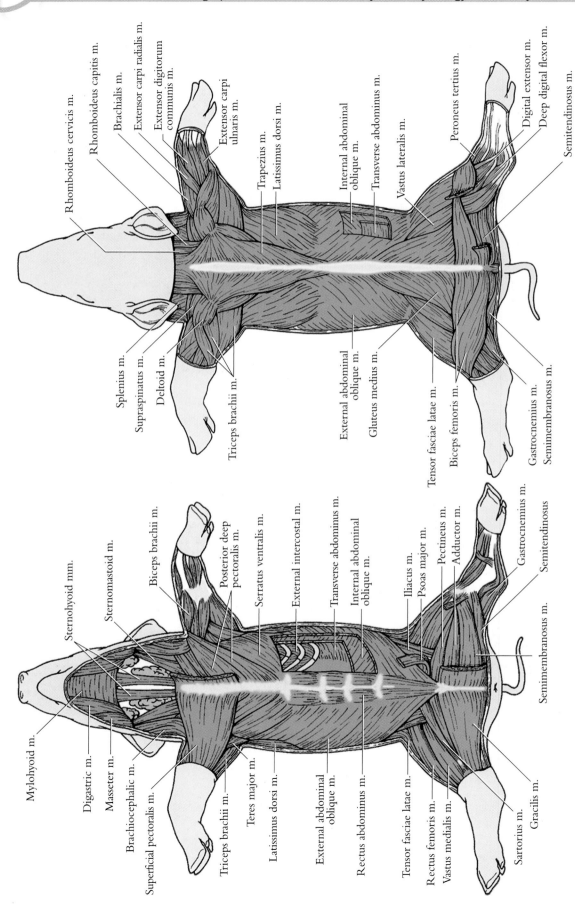

Rhomboideus cervicis m.

Rhomboideus capitis m.

Brachialis m.

Extensor carpi radialis m.

Extensor digitorum communis m.

Extensor carpi uluaris m.

Trapezius m.

Latissimus dorsi m.

Internal abdominal oblique m.

Transverse abdominus m.

Vastus lateralis m.

Peroneus tertius m.

Digital extensor m.

Deep digital flexor m.

Semitendinosus m.

Splenius m.

Supraspinatus m.

Deltoid m.

Triceps brachii m.

External abdominal oblique m.

Gluteus medius m.

Tensor fasciae latae m.

Biceps femoris m.

Gastrocnemius m.

Semimembranosus m.

Figure 20.8
A dorsal view of the muscles of the fetal pig.

Sternohyoid mm.

Sternomastoid m.

Biceps brachii m.

Posterior deep pectoralis m.

Serratus ventralis m.

External intercostal m.

Transverse abdominus m.

Internal abdominal oblique m.

Iliacus m.

Psoas major m.

Pectineus m.

Adductor m.

Gastrocnemius m.

Semitendinosus

Mylohyoid m.

Digastric m.

Masseter m.

Brachiocephalic m.

Superficial pectoralis m.

Triceps brachii m.

Teres major m.

Latissimus dorsi m.

External abdominal oblique m.

Rectus abdominus m.

Tensor fasciae latae m.

Rectus femoris m.

Vastus medialis m.

Sartorius m.

Gracilis m.

Semimembranosus m.

Figure 20.7
A ventral view of the muscles of the fetal pig.

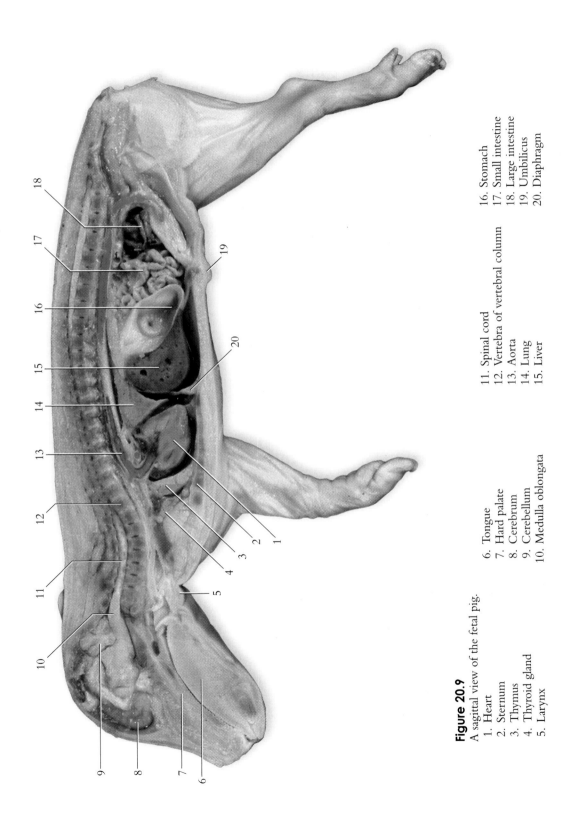

Figure 20.9

A sagittal view of the fetal pig.

1. Heart
2. Sternum
3. Thymus
4. Thyroid gland
5. Larynx
6. Tongue
7. Hard palate
8. Cerebrum
9. Cerebellum
10. Medulla oblongata
11. Spinal cord
12. Vertebra of vertebral column
13. Aorta
14. Lung
15. Liver
16. Stomach
17. Small intestine
18. Large intestine
19. Umbilicus
20. Diaphragm

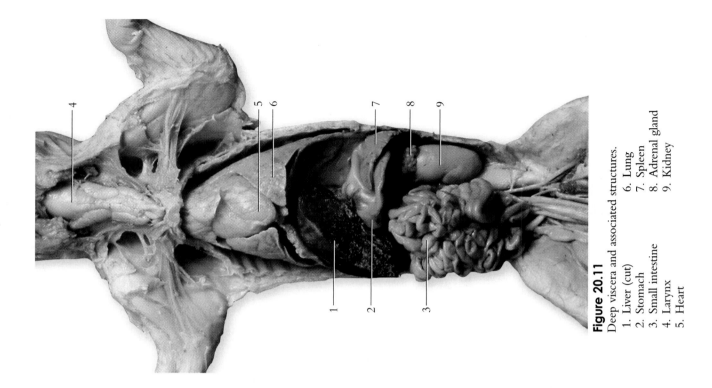

Figure 20.11

Deep viscera and associated structures.

1. Liver (cut)
2. Stomach
3. Small intestine
4. Larynx
5. Heart
6. Lung
7. Spleen
8. Adrenal gland
9. Kidney

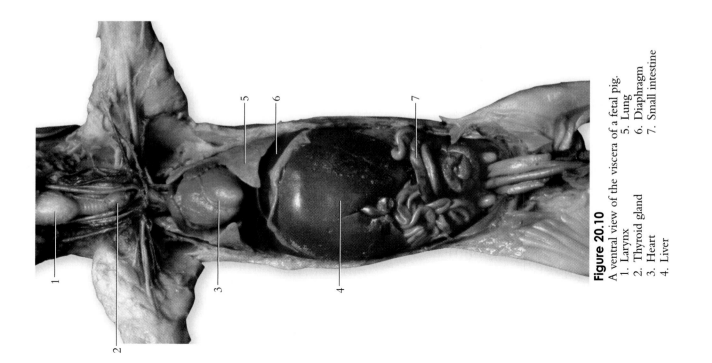

Figure 20.10

A ventral view of the viscera of a fetal pig.

1. Larynx
2. Thyroid gland
3. Heart
4. Liver
5. Lung
6. Diaphragm
7. Small intestine

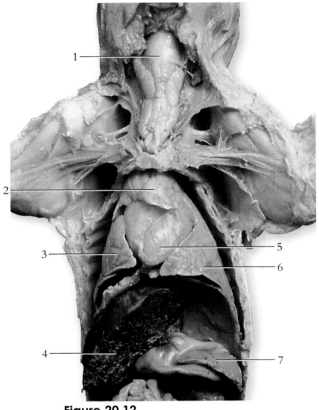

Figure 20.12
Thorax and neck regions of the fetal pig.
1. Larynx
2. Thymus
3. Lung
4. Liver (cut)
5. Heart
6. Lung
7. Spleen (cut)

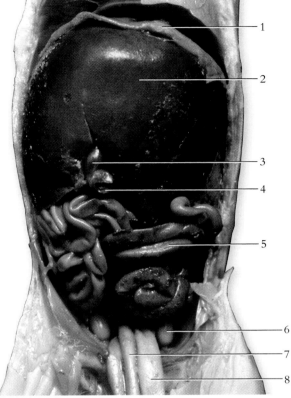

Figure 20.13
A ventral view of the abdominal cavity of a fetal pig.
1. Diaphragm
2. Liver
3. Gallbladder
4. Umbilical vein
5. Small intestine
6. Undescended testis
7. Umbilical artery
8. Urinary bladder

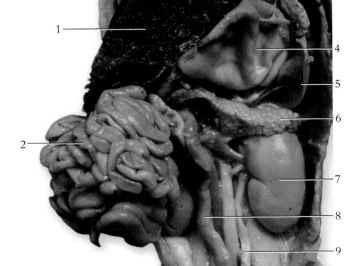

Figure 20.14
Abdominal organs of the fetal pig.
1. Liver (cut)
2. Small intestine
3. Umbilical arteries
4. Stomach (reflected)
5. Spleen
6. Pancreas
7. Kidney
8. Large intestine
9. Ureter
10. Ductus (vas) deferens
11. Urinary bladder

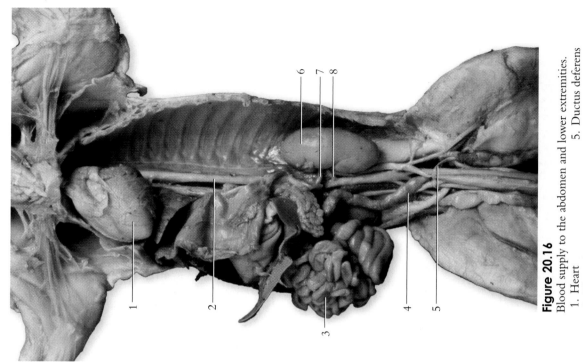

Figure 20.16
Blood supply to the abdomen and lower extremities.
1. Heart
2. Thoracic aorta
3. Small intestine
4. Colon
5. Ductus deferens
6. Kidney
7. Renal artery
8. Renal vein

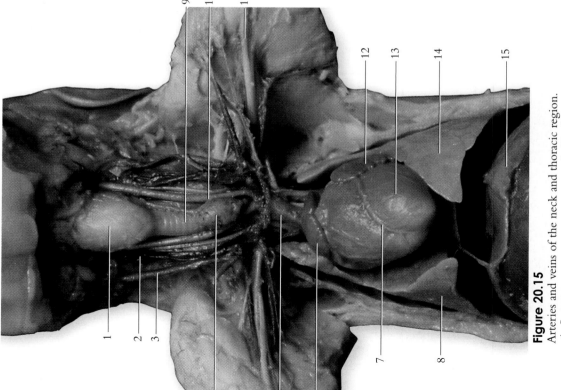

Figure 20.15
Arteries and veins of the neck and thoracic region.
1. Larynx
2. Internal jugular vein
3. External jugular vein
4. Thyroid gland
5. Cranial (superior) vena cava
6. Right auricle
7. Coronary vessels
8. Right lung
9. Trachea
10. Left common carotid artery
11. Axillary artery
12. Left auricle
13. Left ventricle
14. Left lung
15. Diaphragm

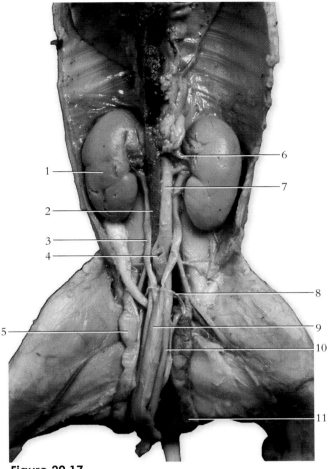

Figure 20.17
Urogenital system of the fetal pig.
1. Kidney
2. Caudal (inferior) vena cava
3. Ureter
4. Rectum (cut)
5. Partially dissected testis
6. Renal vein
7. Descending aorta
8. Ductus deferens
9. Urinary bladder
10. Umbilical artery
11. Epididymis

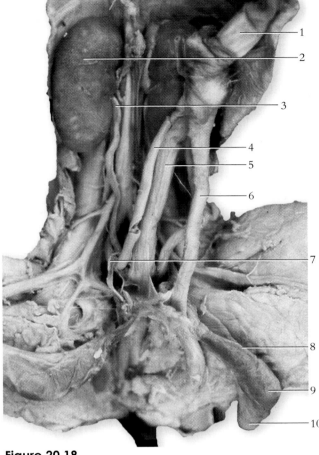

Figure 20.18
Urogenital system of the fetal pig.
1. Umbilical cord
2. Right kidney
3. Ureter
4. Umbilical artery
5. Urinary bladder
6. Penis
7. Vas (ductus) deferens
8. Spermatic cord
9. Right testis
10. Epididymis

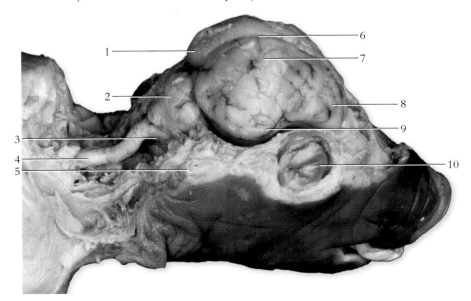

Figure 20.19
General structures of the fetal pig brain. Because the cerebrum is less defined in pigs, the regions are not known as lobes as they are in humans.
1. Occipital region of cerebrum
2. Cerebellum
3. Medulla oblongata
4. Spinal cord
5. External acoustic meatus
6. Longitudinal fissure
7. Parietal region of cerebrum
8. Frontal region of cerebrum
9. Temporal region of cerebrum
10. Eye

㉑ Rat Dissection

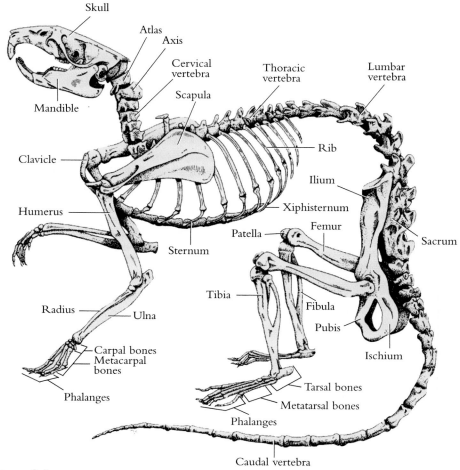

Figure A.1
The skeleton of a rat.

Figure A.2
The laboratory white rat is a captive-raised rodent that is commercially available for biological and medical experiments and research. White rats are also embalmed and available as dissection specimens in biology, vertebrate biology, and general zoology laboratories.

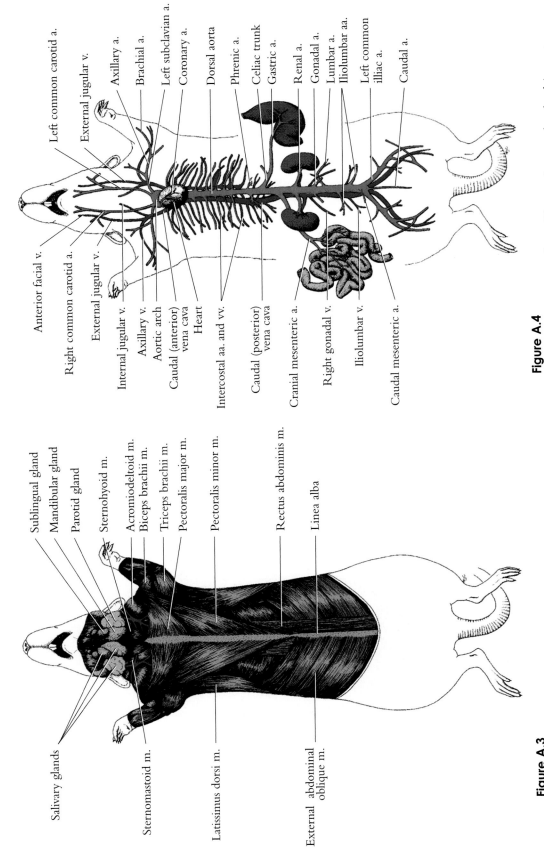

Figure A.4
The circulatory system of a rat. The arteries are colored red (a. = artery, aa. = arteries; v. = vein, vv. = veins).

Left common carotid a.
External jugular v.
Axillary a.
Brachial a.
Left subclavian a.
Coronary a.
Dorsal aorta
Phrenic a.
Celiac trunk
Gastric a.
Renal a.
Gonadal a.
Lumbar a.
Iliolumbar aa.
Left common iliac a.
Caudal a.

Anterior facial v.
Right common carotid a.
External jugular v.
Internal jugular v.
Axillary v.
Aortic arch
Caudal (anterior) vena cava
Heart
Intercostal aa. and vv.
Caudal (posterior) vena cava
Cranial mesenteric a.
Right gonadal v.
Iliolumbar v.
Caudal mesenteric a.

Figure A.3
A ventral view of the rat musculature (m. = muscle).

Sublingual gland
Mandibular gland
Parotid gland
Sternohyoid m.
Acromiodeltoid m.
Biceps brachii m.
Triceps brachii m.
Pectoralis major m.
Pectoralis minor m.
Rectus abdominis m.
Linea alba

Salivary glands
Sternomastoid m.
Latissimus dorsi m.
External abdominal oblique m.

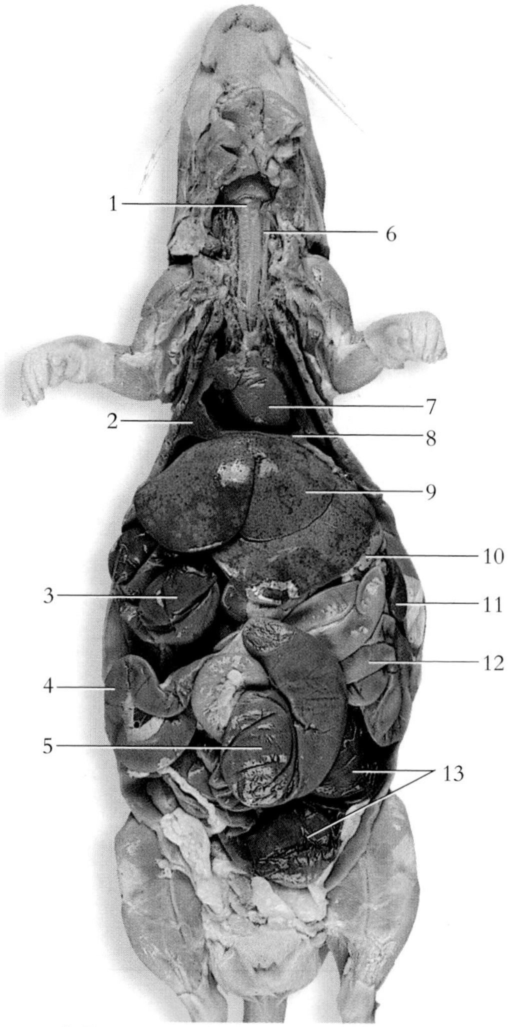

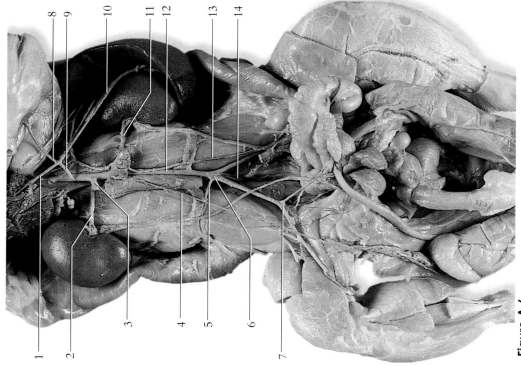

Figure A.6

Abdominal arteries of the rat.

1. Hepatic artery
2. Right renal artery
3. Cranial mesenteric artery
4. Right testicular artery
5. Right iliolumbar artery
6. Caudal mesenteric artery (cut)
7. Right common iliac artery

8. Gastric artery
9. Celiac trunk
10. Splenic artery
11. Left renal artery
12. Abdominal aorta
13. Left testicular artery
14. Middle sacral artery

Figure A.5

A ventral view of the rat viscera.

1. Trachea
2. Right lung
3. Right uterine horn of pregnant female
4. Jejunum
5. Cecum
6. Esophagus
7. Heart

8. Diaphragm (cut)
9. Liver
10. Stomach
11. Spleen
12. Ileum
13. Left uterine horn of pregnant female

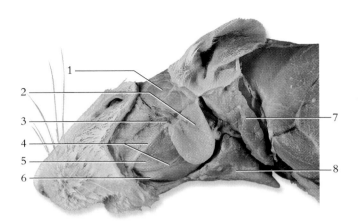

Figure A.7
Head and neck region of the rat.
1. Temporalis muscle
2. Extraorbital lacrimal gland
3. Extraorbital lacrimal duct
4. Facial nerve
5. Masseter muscle
6. Parotid duct
7. Parotid gland
8. Mandibular gland

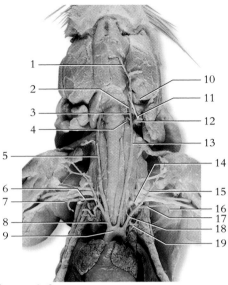

Figure A.8
Arteries of the thoracic and neck regions of the rat.
1. Facial artery
2. Lingual artery
3. External carotid artery
4. Cranial thyroid artery
5. Common carotid artery
6. Axillary artery
7. Brachial artery
8. Brachiocephalic artery
9. Aortic arch
10. External maxillary artery
11. Internal carotid artery
12. Occipital artery
13. Common carotid artery
14. Vertebral artery
15. Cervical trunk
16. Lateral thoracic artery
17. Axillary artery
18. Subclavian artery
19. Internal thoracic artery

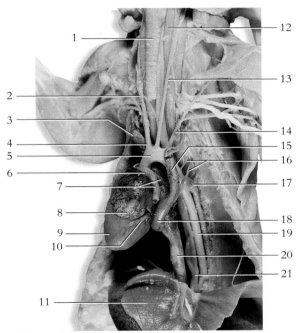

Figure A.9
Reflected rat heart showing the major veins and arteries.
1. Trachea
2. Right common carotid artery
3. Right cranial vena cava
4. Brachiocephalic trunk
5. Aortic arch
6. Pulmonary trunk
7. Left and right pulmonary arteries
8. Left auricle
9. Left ventricle
10. Coronary vein
11. Diaphragm
12. Esophagus
13. Left common carotid artery
14. Left subclavian artery
15. Left cranial vena cava
16. Intercostal artery and vein
17. Azygos vein
18. Coronary sinus
19. Aorta
20. Caudal vena cava
21. Esophagus

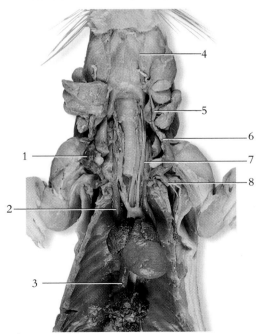

Figure A.10
Veins of the thoracic and neck regions of the rat.
1. Cephalic vein
2. Cranial vena cava
3. Caudal vena cava
4. Linguofacial vein
5. Maxillary vein
6. External jugular vein
7. Internal jugular vein
8. Lateral thoracic vein

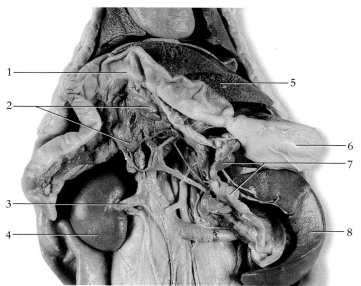

Figure A.11
Abdominal viscera and vessels of the rat.

1. Duodenum
2. Biliary and duodenal parts of pancreas
3. Right renal vein
4. Right kidney
5. Liver (cut)
6. Stomach
7. Gastrosplenic part of pancreas
8. Spleen

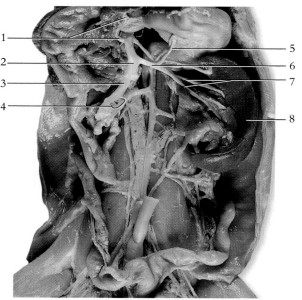

Figure A.12
Branches of the hepatic portal system.

1. Cranial pancreaticoduodenal vein
2. Hepatic portal vein
3. Cranial mesenteric vein
4. Intestinal branches
5. Gastric vein
6. Gastrosplenic vein
7. Splenic branches
8. Spleen

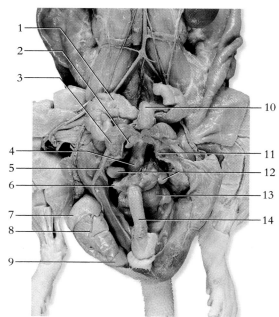

Figure A.13
Urogenital system of the male rat.

1. Vesicular gland
2. Prostate (dorsolateral part)
3. Prostate (ventral part)
4. Urethra in the pelvic canal
5. Ductus (vas) deferens
6. Crus of penis (cut)
7. Head of epididymis
8. Testis
9. Tail of epididymis
10. Urinary bladder
11. Symphysis pubis (cut exposing pelvic canal)
12. Bulbourethral glands
13. Bulbocavernosus muscle
14. Penis

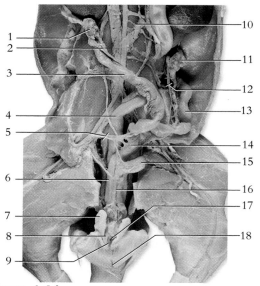

Figure A.14
Urogenital system of the female rat.

1. Ovary
2. Uterine artery and vein
3. Uterine horn
4. Colon
5. Vesicular artery (umbilical artery)
6. Vagina
7. Preputial gland
8. Clitoris
9. Vaginal opening
10. Ovarian artery and vein
11. Ovary
12. Uterine artery and vein
13. Uterine horn
14. Uterine body
15. Urinary bladder
16. Urethra
17. Urethral opening
18. Anus

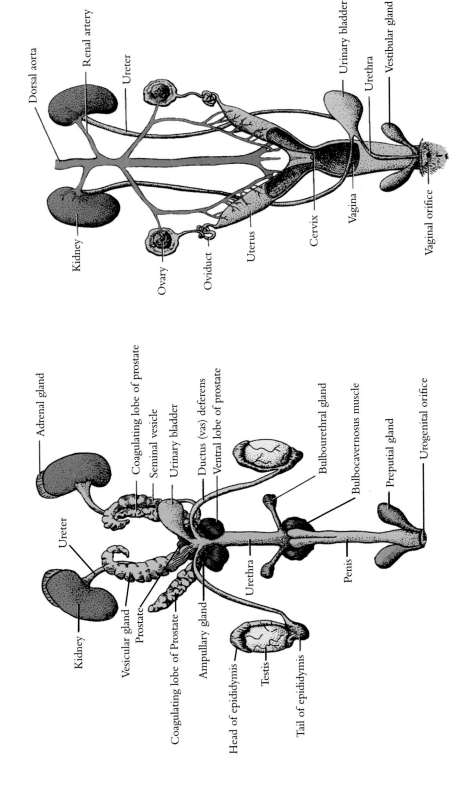

Figure A.16
The urogenital organs of a female rat.

Figure A.15
The urogenital organs of a male rat.

Glossary of Prefixes & Suffixes

Element	Definition and Example	Element	Definition and Example	Element	Definition and Example
a-	absent, deficient or without: atrophy	-coel	swelling, and enlarged space or cavity: blastocoele	end-	within: endoderm
ab-	off, away from: abduct			entero-	intestine: enteritis
abdomin-	abdomen	cephal-	head: cephalis	epi-	upon, in addition: epidermis
-able	capable of: viable	cerebro-	brain: cerebrospinal fluid	erythro-	red: erythrocyte
ac-	toward, to: actin	chol-	bile: cholic	ex-	out of: excise
acou-	hear, acoustic	chondr-	cartilage: chondrocyte	exo-	outside: exocrine
ad-	denoting to, toward: adduct	chrom-	color: chromocyte	extra-	outside of, beyond, in addition: extracellular
af-	movement toward a central point: afferent artery	-cid	destroy: germicide		
		circum-	around: circumduct		
		-cis	cut, kill: excision	fasci-	band: fascia
alba-	pale or white: linea alba	co-	together: copulation	febr-	fever: febrile
-alg	pain: neuralgia	coel-	hollow cavity: coelom	-ferent	bear, carry: efferent arteriole
ambi-	both: ambidextrous	con-	with, together: congenital		
angi-	pertaining to vessel: angiology	contra-	against, opposite: contraception	fiss-	split: fissure
				for-	opening: foramen
ante-	before: antebrachium	corn-	denoting hardness: cornified	-form	shape: fusiform
anti-	against: anticoagulant	corp-	body: corpus	gastro-	stomach: gastrointestinal
aqua-	water: aqueous	crypt-	hidden: cryptorchism	-gen	an agent that produces or originates: pathogen
archi-	to be first: archeteron	cyan-	blue color: cyanosis		
arthri-	joint: arthritis	cysti-	sac or bladder: cystoscope	-genic	produced from, producing: carcinogenic
-asis	condition or state of: homeostasis	cyto-	cell: cytology	gloss-	tongue: glossopharyngeal
aud-	pertaining to ear: auditory			glyco-	sugar: glycosuria
		de-	down, from: descent	-gram	a record, recording: myogram
auto-	self: autolysis	derm-	skin: dermatology		
		di-	two: diarthrotic	gran-	grain, particle: agranulocyte
bi-	two: bipedal	dipl-	double: diploid		
bio-	life: biology	dis-	apart, away from: disarticulate	-graph	instrument for recording: electrocardiograph
blast-	generative or germ bud: osteoblast	duct-	lead, conduct: ductus deferens	grav-	heavy: gravid
brachi-	arm: brachialis			gyn-	female sex: gynocology
brachy-	short: brachydont	dur-	hard: dura mater		
brady-	slow: bradycardia	-dynia	pain	hema(o)-	blood: hematology
bucc-	cheek: buccal cavity	dys-	bad, difficult, painful: dysentery	haplo-	simple or single: haploid
				hemi-	half: hemiplegia
cac-	bad, ill: cachexia			hepat-	liver: hepatic portal
calci-	stone: calculus	e-	out, from: eccrine	hetero-	other, different: heterosexual
capit-	head: capitis	ecto-	outside, outer, external: ectoderm		
carcin-	cancer: carcinogenic			histo-	webb, tissue: histology
cardi-	heart: cardiac	-ectomy	surgical removal: tonsillectomy	holo-	whole, entire: holocrine
caud-	tail: cauda equina	ede-	swelling: edema	homo-	same, alike: homologous
cata-	lower, under, against: catabolism	-emia	pertaining to a condition of the blood: lipemia	hydro-	water: hydrocoel
				hyper-	beyond, above, excessive: hypertension

Element	Definition and Example	Element	Definition and Example	Element	Definition and Example
hypo-	under, below: hypoglycemia	necro-	corpse, dead: necrosis	-pnea	to breathe: apnea
		nephro-	kidney: nephritis	pneumato-	breathing: pneumonia
-ia	state or condition: hypoglycemia	neuro-	nerve: neurolemma	pod-	foot: podiatry
		noto-	back: notochord	-poieis	formation of: hematopoiesis
-iatrics	medical specialties: pediatrics	ob-	against, toward, in front of: obturator	poly-	many, much: polyploid
idio-	self, separate, distinct: idiopathic			post-	after, behind: post natal
ilio-	ilium: iliosacral	oc-	against: occlusion	pre-	before in time or place: prenatal
infra-	beneath: infraspinatus	-oid	resembling, likeness: sigmoid	prim-	first: primitive
inter-	among, between	oligo-	few, small: oligodendrocyte	pro-	before in time or place: prosect
intra-	inside, within: intracellular	-oma	tumor: lymphoma	proct-	anus: proctology
-ion	process: acromion	oo-	egg: oocyte	pseudo-	false: pseudostratified
iso-	equal, like: isotonic	or-	mouth: oral	psycho-	mental: psychology
-ism	condition or state: rheumatism	orchi-	testicles: cryptorchidism	pyo-	pus: pyoculture
		-ory	pertaining to: sensory		
-itis	inflammation: meningitis	osteo-	bone: osteoblast	quad-	fourfold: quadriceps femoris
		-ose	full of: adipose		
		oto	ear: otolith		
labi-	lip: labium majus	ovo-	egg: ovum	re-	back, again: repolarization
lacri-	tears: nasolacrimal			rect-	straight: rectus abdominis
later-	side: lateral	para-	give birth to, bear: parturition	reno-	kidney: renal
-logy	science of: morphology			rete-	network: retina
-lysis	solution, dissolve: hemolysis	para-	near, beyond, beside: paranasal	retro-	backward: retroperitoneal
macro-	large, great: macrophage	path-	disease, that which undergoes sickness: pathology	rhin-	nose: rhinitis
mal-	bad, abnormal, disorder: malignant			-rrhage	excessive flow: hemorrhage
medi-	middle: medial	-pathy	abnormality, disease: neuropathy	-rrhea	flow, or discharge: diarrhea
mega-	great, large: mega karyocyte	ped-	children: pediatrician		
		pen-	need, lack: penicillin		
meso-	middle or moderate: mesoderm	-penia	deficiency: thrombocytopenia	sanguin-	blood: sanguiferous
meta-	after, beyond: metatarsal			sarc-	flesh: sarcoplasm
micro-	small: microtome	per-	through: percutaneous	-scope	instrument for examination of a part: stethoscope
mito-	thread: mitosis	peri-	near, around: pericardium		
mono-	alone, one, single: monocyte	phag-	to eat: phagocyte	-sect	cut: dissect
mons-	mountain: mons pubis	-phil	have an affinity for: neutrophil	semi-	half: semilunar
morph-	form, shape: morphology	phlebo-	vein: phlebitis	serrate-	saw-edged: serratus anterior
multi-	many, much: multinuclear	-phobe	abnormal fear, dread: hydrophobia	-sis	state or condition: dialysis
myo-	muscle: myology	-plasty	reconstruction of: rhinoplasty	steno-	narrow: stenohaline
narc-	numbness, stupor: narcotic	platy-	flat, side: platysma	-stomy	surgical opening: tracheotomy
neo-	new, young: neonatal	-plegia	stroke, paralysis: paraplegia	sub-	under, beneath, below: subcutaneous

Element	Definition and Example
super–	above, beyond, upper: superficial
supra–	above, over: suprarenal
syn (sym)	together, joined, with: synapse
tachy–	swift, rapid: tachometer
tele–	far: telencephalon
tens–	stretch: tensor fascia lata
tetra–	four: tetrad
therm–	heat: thermogram
thorac–	chest: thoracic cavity
thrombo–	lump, clot: thrombocyte
–tomy	cut: appendectomy
tox–	poison: toxic
tract–	draw, drag: traction
trans–	across, over: transfuse
tri–	three: trigone
trich–	hair: trichology
–trophy	a state relating to nutrition: hypertrophy
–tropic	turning toward, changing: gonadotropic
ultra–	beyond, excess: ultrasonic
uni–	one: unicellular
uro–	urine, urinary organs or tract: uroscope
–uria	urine: polyuria
vas–	vessel: vasoconstriction
vermi–	worm: vermiform
viscer–	organ: visceral
vit–	life: vitamin
zoo–	animal: zoology
zygo–	union, join: zygote

abdomen (ab-do´men): the portion of the trunk located between the diaphragm and the pelvis; contains the abdominal cavity and its visceral organs.

abduction (ab-duk´shun): a movement away from the axis or midline of the body; opposite of adduction; a movement of a digit away from the axis of a limb.

acapnia (ah-kap´ne-ah): a decrease in normal amount of CO_2 in the blood.

accommodation (ah-kom-o-da´shun): a change in the shape of the lens of the eye so that vision is more acute; the focusing for various distances.

acetone (as´e-tone): an organic compound that may be present in the urine of diabetics; also called ketone bodies.

Achilles tendon (ah-kil´ez): see *tendo calcaneus*.

acidosis (as-i-do´sis): a disorder of body chemistry in which the alkaline substances of the blood are reduced below normal.

actin (ak´tin): a protein in muscle fibers that together with myosin is responsible for contraction.

acoustic (ah-koos´tik): referring to sound or the sense of hearing.

adduction (ah-duk´shun): a movement toward the axis or midline of the body; opposite of abduction; a movement of a digit toward the axis of a limb.

adenohypophysis (ad´e-no-hi-pof´i-sis): anterior pituitary gland.

adenoid (ad´e-noid): paired lymphoid structures in the nasopharynx; also called pharyngeal tonsils.

adenosine triphosphate (ATP) (ah-den´o-sen tri-fos´fate): a chemical compound that provides energy for cellular use.

adipose (ad´e-pose): fat, or fat-containing, such as adipose tissue.

adrenal glands (ah-dre´nal): endocrine glands; one superior to each kidney; also called *suprarenal glands*.

aerobic (a-er-o´bik): requiring free O_2 for growth and metabolism as in the case of certain bacteria called *aerobes*.

allantois (ah-lan´to-is): an extraembryonic membranous sac that forms blood cells and gives rise to the fetal umbilical arteries and vein. It also contributes to the formation of the urinary bladder.

alveolus (al-ve´o-lus): an individual aircapsule within the lung. Alveoli are the basic functional units of respiration. Also, the socket that secures a tooth.

amnion (am´ne-on): a membrane that surrounds the fetus to contain the amniotic fluid.

amphiarthrosis (am´fe-ar-thro´sis): a slightly moveable joint in a functional classification of joints.

anatomical position (an´ah-tom´e-kal): an erect body stance with the eyes directed forward, the arms at the sides, and the palms of the hands facing forward.

anatomy (ah-nat´o-me): the branch of science concerned with the structure of the body and the relationship of its organs.

antebrachium (an´te-bra´ke-um): the forearm.

anterior (ventral) (an-te´re-or): toward the front; the opposite of *posterior (dorsal)*.

antigen (an´te-jen): a substance which causes cells to produce antibodies.

anus (a´nus): the terminal end of the GI tract, opening of the anal canal.

aorta (a-or´tah): the major systemic vessel of the arterial portion of the circulatory system, emerging from the left ventricle.

apocrine gland (ap´o-krin): a type of sweat gland that functions in evaporative cooling.

appendix (ah-pen´diks): a short pouch that attaches to the cecum.

aqueous humor (a´kwe-us hu´mor): the watery fluid that fills the anterior and posterior chambers of the eye.

arachnoid mater (ah-rak´noid): the weblike middle covering (meninx) of the central nervous system.

arbor vitae (ar´bor vi´tah): the branching arrangement of white matter within the cerebellum.

areola (ah-re´o-lah): the pigmented ring around the nipple.

artery (ar´ter-e): a blood vessel that carries blood away from the heart.

articular cartilage (ar-tik´u-lar ker´ti-lij): a hyaline cartilaginous covering over the articulating surface of bones of synovial joints.

ascending colon (ko´lon): the portion of the large intestine between the cecum and the right colic (hepatic) flexure.

atom (at´om): the smallest unit of an element that can exist and still have the properties of the element; collectively, atoms form molecules in a compound.

atrium (a´tre-um): either of two superior chambers of the heart that receive venous blood.

atrophy (at´ro-fe): a wasting away or decrease in size of a cell or organ.

auditory tube (aw´di-to´re): a narrow canal that connects the middle ear chamber to the pharynx; also called the *eustachian canal*.

autonomic (aw-to-nom´ik): self-governing; pertaining to the division of the nervous system which controls involuntary activities.

axilla (ak-sil´ah): the depressed hollow under the arm; the armpit.

axon (ak´son): The elongated process of a neuron (nerve cell) that transmits an impulse away from the cell body.

basement membrane: a thin sheet of extracellular substance to which the basal surfaces of membranous epithelial cells are attached.

basophil (ba´so-fil): a granular leukocyte that readily stains with basophilic dye.

belly: the thickest circumference of a skeletal muscle.

benign: (be-nine´): nonmalignant; a confined tumor.

blastula (blas´tu-lah): an early stage of prenatal development between the morula and embryonic stages.

blood: the fluid connective tissue that circulates through the cardiovascular system to transport substances throughout the body.

bolus (bo´lus): a moistened mass of food that is swallowed from the oral cavity into the pharynx.

bone: an organ composed of solid, rigid connective tissue, forming a component of the skeletal system.

Bowman's capsule (bo´manz kap´sul): see *glomerular capsule.*

brain: the enlarged superior portion of the central nervous system, located in the cranial cavity of the skull.

brain stem: the portion of the brain consisting of the medulla oblongata, pons, and midbrain.

bronchial tree (brong´ke-al): the bronchi and their branching bronchioles.

bronchiole (brong´ke-ol): a small division of a bronchus within the lung.

bronchus (bron´kus): a branch of the trachea that leads to a lung.

buccal cavity (buk´al): the mouth, or oral cavity.

bursa (ber´sah): a saclike structure filled with synovial fluid, which occurs around joints.

buttock (but´ok):the rump or fleshy mass on the posterior aspect of the lower trunk, formed primarily by the gluteal mucles.

calorie (kal´o-re): the unit of heat required to raise the temperature of one gram of water one degree centigrade.

calyx (ka´liks): a cup-shaped portion of the renal pelvis that encircles a renal papilla.

cancellous bone (kan´se-lus): spongy bone; bone tissue with a latticelike structure.

capillary (kap´i-lar´e): a microscopic blood vessel that connects an arteriole and a venule; the functional unit of the circulatory system.

carcinogenic (kar-si-no-jen´ik): stimulating or causing the growth of a malignant tumor, or cancer.

carpus (kar´pus): the proximal portion of the hand that contains the eight carpal bones.

cartilage (kar´ti´lij): a type of connective tissue with a solid elastic matrix.

caudal (kaw´dal): referring to a position more toward the tail.

cecum (se´kum): the pouchlike portion of the large intestine to which the ileum of the small intestine is attached.

cell: the structural and functional unit of an organism; the smallest structure capable of performing all the functions necessary for life.

central nervous system (CNS): the brain and the spinal cord.

centrosome (sen´tro-som): a dense body near the nucleus of a cell that contains a pair of centrioles.

cerebellum (ser´e-bel´um): the portion of the brain concerned with the coordination of movements and equilibrium.

cerebrospinal fluid (ser´e-bro-spi´nal): a fluid that buoys and cushions the central nervous system.

cerebrum (ser´e-brum): the largest portion of the brain, composed of the right and left hemispheres.

cervical (ser´vi-kal): pertaining to the neck or a necklike portion of an organ.

choanae (ko-a´na): the two posterior openings from the nasal cavity into the nasopharynx.

cholesterol (ko-les´ter-ol): an organic fat-like compound found in animal fat, bile, blood, liver, and other parts of the body.

chondrocyte (kon´dro-site): a cartilage cell.

chorion (ko´re-on): An extraembryonic membrane that participates in the formation of the placenta.

choroid (ko´roid): the vascular, pigmented middle layer of the wall of the eye.

chromosome (kro´mo-som): structure in the nucleus that contains the genes for genetic expression.

chyme (kime): The mass of partially digested food that passes from the stomach into the duodenum of the small intestine.

cilia (sil´e-ah): microscopic, hairlike processes that move in a wavelike manner on the exposed surfaces of certain epithelial cells.

ciliary body (sil´e-er´e): a portion of the choroid layer of the eye that secretes aqueous humor and contains the ciliary muscle.

circumduction (ser´kum-duk´shun): a conelike movement of a body part, such that the distal end moves in a circle while the proximal portion remains relatively stable.

clitoris (kli´to-ris): a small, erectile structure in the vulva of the female.

cochlea (kok´le-ah): the spiral portion of the inner ear that contains the spiral organ (organ of Corti).

ceolom (se´lom): the abdominal cavity.

colon (ko´lon): the first portion of the large intestine.

common bile duct: a tube that is formed by the union of the hepatic duct and cystic duct, transports bile to the duodenum.

compact bone: tightly packed bone that is superficial to spongy bone; also called *dense bone.*

condyle (kon´dile): a rounded process at the end of a long bone that forms an articulation.

connective tissue: one of the four basic tissue types within the body. It is a binding and supportive tissue with abundant matrix.

cornea (kor´ne-ah): the transparent convex, anterior portion of the outer layer of the eye.

cortex (kor´teks): the outer layer of an organ such as the convoluted cerebrum, adrenal gland, or kidney.

costal cartilage (kos´tal): the cartilage that connects the ribs to the sternum.

cranial (kra´ne-al): pertaining to the cranium.

cranial nerve: one of twelve pairs of nerves that arise from the inferior surface of the brain.

dentin (den´tine): the principal substance of a tooth, covered by enamel over the crown and by cementum on the root.

dermis (der´mis): the second, or deep, layer of skin beneath the epidermis.

descending colon: the segment of the large intestine that descends on the left side from the level of the spleen to the level of the left iliac crest.

diaphragm (di´ah-fram): a flat dome of muscle and connective tissue that separates the thoracic and abdominal cavities.

diaphysis (di-af´i-sis): the shaft of a long bone.

diastole (di-as´to-le): the sequence of the cardiac cycle during which the ventricular heart chamber wall is relaxed.

diarthrosis (di´ar-thro´sis): a freely movable joint.

distal (dis´tal): away from the midline or origin; the opposite of *proximal.*

dorsal (dor´sal): pertaining to the back or posterior portion of a body part; the opposite of *ventral.*

ductus deferens (duk'tus def'er-enz): a tube that carries spermatozoa from the epididymis to the ejaculatory duct: also called the *vas deferens* or *seminal duct*.

duodenum (du'o-num): the first portion of the small intestine.

dura mater (du'rah ma'ter): the outermost meninx covering the central nervous system.

eccrine gland (ek'rin): a sweat gland that functions in body cooling.

ectoderm (ek'to-derm): the outermost of the three primary embryonic germ layers.

edema (e-de'mah): an excessive retention of fluid in the body tissues.

effector (ef-fek'tor): an organ such as a gland or muscle that responds to motor stimulation.

efferent (ef'er-ent): conveying away from the center of an organ or structure.

ejaculation (e-jak'u-la'shun): the discharge of semen from the male urethra during climax.

electrocardiogram (e-lek'tro-kar'de-o-gram'): a recording of the electrical activity that accompanies the cardiac cycle; also called ECG or EKG.

electroencephalogram (e-lek'tro-en-sef'ah-lo-gram): a recording of the brain wave pattern; also called EEG.

electromyogram (e-lek'tro-mi'o-gram): a recording of the activity of a muscle during contraction: also called EMG.

electrolyte (e-lek'tro-lite): a solution that conducts electricity by means of charged ions.

electron (e-lek'tron): the unit of negative electricity.

enamel (en-am'el): the outer, dense substance covering the crown of a tooth.

endocardium (en'do-kar'de-um): the fibrous lining of the heart chambers and valves.

endochondral bone (en'do-kon'dral): bones that form as hyaline cartilage models first and then are ossified.

endocrine gland (en'do-krine): hormone producing gland that is part of the endocrine system.

endoderm (en'do-derm): the innermost of the three primary germ layers of an embryo.

endometrium (en'do-me'tre-um): the inner lining of the uterus.

endothelium (en'do-the'le-um): the layer of epithelial tissue that forms the thin inner lining of blood vessels and heart chambers.

eosinophil (e'o-sin'o-fil): a type of white blood cell that becomes stained by acidic eosin dye; constitutes about 2%–4% of the white blood cells.

epicardium (ep'i-kar'de-um): the thin, outer layer of the heart: also called the *visceral pericardium*.

epidermis (ep'i-der'mis): the outermost layer of the skin, composed of stratified squamous epithelium.

epididymis (ep'i-did'i-mis): a coiled tube located along the posterior border of the testis; stores spermatozoa and discharges them during ejaculation.

epidural space (ep'i-du'ral): a space between the spinal dura mater and the bone of the vertebral canal.

epiglottis (ep'i-glot'is): a leaflike structure positioned on top of the larynx that covers the glottis during swallowing.

epinephrine (ep'i-nef'rin): a hormone secreted from the adrenal medulla resulting in action similar to those from sympathetic nervous system stimulation; also called *adrenaline*.

epiphyseal plate (ep'i-fize-al): a cartilaginous layer located between the epiphysis and diaphysis of a long bone and functions in longitudinal bone growth.

epiphysis (e-pif'i-sis): the end segment of a long bone, distinct in early life but later becoming part of the larger bone.

epithelial tissue (ep'i-the'le-al): one of the four basic tissue types; the type of tissue that covers or lines all exposed body surfaces.

erythrocyte (e-rith'ro-site): a red blood cell.

esophagus (e-sof'ah-gus): a tubular organ of the GI tract that leads from the pharynx to the stomach.

estrogen (es'tro-jen): female sex hormone secreted from the ovarian (Graafian) follicle.

eustachian canal (u-sta'ke-an): see *auditory tube*.

excretion (eks-kre'shun): discharging waste material.

exocrine gland (ek'so-krin): a gland that secretes its product to an epithelial surface, directly or through ducts.

expiration (ek'spi-ra'shun): the process of expelling air from the lungs through breathing out; also called *exhalation*.

extension (ek-sten'shun): a movement that increases the angle between two bones of a joint.

external ear: the outer portion of the ear, consisting of the auricle (pinna), external auditory canal.

extracellular (esk-trah-sel'u-lar): outside a cell or cells.

extrinsic (eks-trin'sik): pertaining to an outside or external origin.

facet (fas'et): a small, smooth surface of a bone where articulation occurs.

fallopian tube (fal-lo'pe-an): see *uterine tube*.

fascia (fash'e-ah): a tough sheet of fibrous connective tissue binding the skin to underlying muscles or supporting and separating muscle.

fasciculus (fah-sik'u-lus): a bundle of muscle or nerve fibers.

feces (fe'sez): waste material expelled from the GI tract during defecation, composed of food residue, bacteria, and secretions; also called *stool*.

fetus (fe'tus): the unborn offspring during the last stage of prenatal development.

filtration (fil-tra'shun): the passage of a liquid through a filter or a membrane.

fimbriae (fim'bre-e): fringelike extensions from the borders of the open end of the uterine tube.

fissure (fish'ure): a groove or narrow cleft that separates two parts of an organ.

flexion (flek'shun): a movement that decreases the angle between two bones of a joint; opposite of extension.

fontanel (fon'tah-nel): a membranous-covered region on the skull of a fetus or baby where ossification has not yet occurred: also called a *soft spot*.

foot: the terminal portion of the lower extremity, consisting of the tarsus, metatarsus, and digits.

foramen (fo-ra´men): an opening in an anatomical structure for the passage of a blood vessel or a nerve.

foramen ovale (o-val´e): the opening through the interatrial septum of the fetal heart.

fossa (fos´ah): a depressed area, usually on a bone.

fourth ventricle (ven´tri-k´l): a cavity within the brain containing cerebrospinal fluid.

fovea centralis (fo´ve-ah sen´tra´lis): a depression on the macula lutea of the eye where only cones are located, which is the area of keenest vision.

gallbladder: a pouchlike organ, attached to the inferior side of the liver, which stores and concentrates bile.

gamete (gam´ete): a haploid sex cell, sperm or egg.

gamma globulins (gam´mah glob´u-lins): protein substances often found in immune serums that act as antibodies.

ganglion (gang´gle-on): an aggregation of nerve cell bodies outside the central nervous system.

gastrointestinal tract (gas´tro-in-tes´tin-al): the tubular portion of the digestive system that includes the stomach and the small and large intestines; also called the *GI tract*.

gene (jene): one of the biologic units of heredity; parts of the DNA molecule located in a definite position on a certain chromosome.

genetics (je-net´iks): the study of heredity.

gingiva (jin-ji´vah): the fleshy covering over the mandible and maxilla through which the teeth protrude within the mouth; also called the *gum*.

gland: an organ that produces a specific substance or secretion.

glans penis (glanz pe´nis): the enlarged, distal end of the penis.

glomerular capsule (glo-mer´u-lar): the double-walled proximal portion of a renal tubule that encloses the glomerulus of a *nephron*; also called *Bowman's capsule*.

glomerulus (glo-mer´u´lus): a coiled tuft of capillaries that is surrounded by the glomerular capsule and filters urine from the blood.

glottis (glot´is): a slitlike opening into the larynx, positioned between the vocal folds.

glycogen (gli´ko-jen): the principal storage carbohydrate in animals. It is stored primarily in the liver and is made available as glucose when needed by the body cells.

goblet cell: a unicellular gland within columnar epithelia that secretes mucus.

gonad (go´nad): a reproductive organ, testis or ovary, that produces gametes and sex hormones.

gray matter: the portion of the central nervous system that is composed of nonmyelinated nervous tissue.

greater omentum (o-men´tum): a double-layered peritoneal membrane that originates on the greater curvature of the stomach and extends over the abdominal viscera.

gut: pertaining to the intestine; generally a developmental term.

gyrus (ji´rus): a convoluted elevation or ridge.

hair: an epidermal structure consisting of keratinized dead cells that have been pushed up from a dividing basal layer.

hair cells: specialized receptor nerve endings for responding to sensations, such as in the spiral organ of the inner ear.

hair follicle (fol´li-k´l): a tubular depression in the skin in which a hair develops.

hand: the terminal portion of the upper extremity, consisting of the carpus, metacarpus, and digits.

hard palate (pal´at): the bony partition between the oral and nasal cavities, formed by the maxillae and palatine bones.

haustra (haws´trh): sacculations or pouches of the colon.

haversian system (ha-ver´shan): see *osteon*.

heart: a muscular, pumping organ positioned in the thoracic cavity.

hematocrit (he-mat´o-krit): the volume percentage of red blood cells in whole blood.

hemoglobin (he´mo-glo´bin): the pigment of red blood cells that transports O_2 and CO_2.

hemopoiesis (hem´ah-poi-e´sis): production of red blood cells.

hepatic portal circulation (por´tal): the return of venous blood from the digestive organs and spleen through a capillary network within the liver before draining into the heart.

heredity (re-red´i-te): the transmission of certain characteristics, or traits, from parents to offspring, via the genes.

hiatus (hi-a´tus): an opening or fissure.

hilum (hi´lum): a concave or depressed area where vessels or nerves enter or exit an organ.

histology (his-tol´o-je): microscopic anatomy of the structure and function of tissues.

homeostasis (ho-me-o-sta´sis): a consistency and uniformity of the internal body environment which maintains normal body function.

hormone (hor´mone): a chemical substance that is produced in an endocrine gland and secreted into the bloodstream to cause an effect in a specific target organ.

hyaline cartilage (hi´ah-line): the most common kind of cartilage in the body, occurring at the articular ends of bones, in the trachea, and within the nose, and forms the precursor to most of the bones of the skeleton.

hymen (hi´men): a developmental remnant (vestige) of membranous tissue that partially covers the vaginal opening.

hyperextension (hi´per-ek-sten´shun): extension beyond the normal anatomical position of 180°.

hypothalamus (hi´po-thal´ah-mus): a structure within the brain below the thalamus, which functions as an autonomic center and regulates the pituitary gland.

ileocecal valve (il´e-o-se´kal): a specialization of the mucosa at the junction of the small and large intestine that forms a one-way passage and prevents the backflow of food materials.

ileum (il´e-um): the terminal portion of the small intestine between the jejunum and cecum.

inferior vena cava (ve´nah ka´vah): a systemic vein that collects blood from the body regions inferior to the level of the heart and returns it to the right atrium.

inguinal (ing´gwi-nal): pertaining to the groin region.

inguinal canal: the passage in the abdominal wall through which a testis descends into the scrotum.

insertion: the more movable attachment of a muscle, usually more distal in location.

inspiration (in'spi-ra'shun): the act of breathing air into the alveoli of the lungs; also called *inhalation.*

integument (in-teg'u-ment): pertaining to the skin.

internal ear: the innermost portion or chamber of the ear, containing the cochlea and the vestibular organs.

interstitial (in-ter-stish'al): pertaining to spaces or structures between the functioning active tissue of any organ.

intracellular (in-trah-sel'u-lar): within the cell itself.

intervertebral disc (in'ter-ver'te-bral): a pad of fibrocartilage between the bodies of adjacent vertebrae.

intestinal gland (in-tes'ti-nal): a simple tubular digestive gland that opens onto the surface of the intestinal mucosa and secretes digestive enzymes; also called *crypt of Lieberkuhn.*

intrinsic (in-trin'sik): situated in or pertaining to an internal organ.

iris (i'ris): the pigmented, vascular tunic portion of the eye that surounds the pupil and regulates its diameter.

islets of Langerhans (i'lets of lahng'er-hanz): see *pancreatic islets.*

isotope (i'so-tope): a chemical element that has the same atomic number as another but a different atomic weight.

isthmus (is'mus): a narrow neck or portion of tissue connecting two structures.

jejunum (je-joo'num): the middle portion of the small intestine, located between the duodenum and the ileum.

joint capsule (kap'sule): the fibrous tissue that encloses the joint cavity of a synovial joint.

jugular (jug'u-lar): pertaining to the veins of the neck which drain the areas supplied by the carotid arteries.

karyotype (kar'e-o-tip): the arrangement of chromosomes that is characteristic of the species or of a certain individual.

keratin (ker'ah-tin): an insoluble protein present in the epidermis and in epidermal derivatives such as hair and nails.

kidney (kid'ne): one of the paired organs of the urinary system that contains nephrons and filters wastes from the blood in the formation of urine.

labia majora (la'be-ah ma-jor'ah): a portion of the external genitalia of a female, consisting of two longitudinal folds of skin extending downward and backward from the mons pubis.

labia minora (ma-nor'ah): two small folds of skin, devoid of hair and sweat glands, lying between the labia majora of the external genitalia of a female.

lacrimal gland (lak'ri-mal): a tear-secreting gland, located on the superior lateral portion of the eyeball underneath the upper eyelid.

lactation (lak-ta'shun): the production and secretion of milk by the mammary glands.

lacteal (lak'te-al): a small lymphatic duct within a villus of the small intestine.

lacuna (lah-ku'nah): a hollow chamber that houses an osteocyte in mature bone tissue or a chondrocyte in cartilage tissue.

lamella (lah-mel'ah): a concentric ring of matrix surrounding the central canal in an osteon of mature bone tissue.

large intestine: the last major portion of the GI tract, consisting of the cecum, colon, rectum, and anal canal.

larynx (lar'inks): the structure located between the pharynx and trachea that houses the vocal folds (cords); commonly called the *voice box.*

lens (lenz): a transparent refractive structure of the eye, derived from ectoderm and positioned posterior to the pupil and iris.

leukocyte (lu'ko-site): a white blood cell; also spelled *leucocyte.*

ligament (lig'ah-ment): a fibrous band or cord of connective tissue that binds bone to bone to strengthen and provide support to the joint; also may support viscera.

limbic system (lim'bik): a portion of the brain concerned with emotions and autonomic activity.

linea alba (lin'e-ah al'bah): a fibrous band extending down the anterior medial portion of the abdominal wall.

locus (lo'kus): the specific location or site of a gene within the chromosome.

lumbar (lum'bar): pertaining to the region of the loins.

lumen (lu'men): the space within a tubular structure through which a substance passes.

lung: one of the two major organs of respiration within the thoracic cavity.

lymph (limf): a clear fluid that flows through lymphatic vessels.

lymph node: a small, ovoid mass located along the course of lymph vessels.

lymphocyte (lim'fo-site): a type of white blood cell characterized by a granular cytoplasm.

macula lutea (mak'u-lah lu'te-ah): a depression in the retina that contains the fovea centralis, the area of keenest vision.

malignant (mah-lig'nant): a disorder that becomes worse and eventually causes death, as in cancer.

malnutrition (mal-nu-trish'un): any abnormal assimilation of food; receiving insufficient nutrients.

mammary gland (mam'er-e): the gland of the female breast responsible for lactation and nourishment of the young.

marrow (mar'o): the soft vascular tissue that occupies the inner cavity of certain bones and produces blood cells.

matrix (ma'triks): the intercellular substance of a tissue.

meatus (me-a'tus): an opening or passageway into a structure.

mediastinum (me'de-as-ti'num): the partition in the center of the thorax between the two pleural cavities.

medulla (me-dul'ah): the center portion of an organ.

medulla oblongata (ob'long-ga'tah): a portion of the brain stem between the pons and the spinal cord.

medullary cavity (med'u-lar'e): the hollow center of the diaphysis of a long bone, occupied by marrow.

meiosis (mi-o'sis): cell division by which gametes, or haploid sex cells, are formed.

melanocyte (mel'ah-no-site): a pigment-producing cell in the deepest epidermal layer of the skin.

membranous bone (mem'brah-nus): bone that forms from membranous connective tissue rather than from cartilage.

menarche (me-nar'ke): the first menstrual discharge.

meninges (me-nin'jez): a group of three fibrous membranes that cover the central nervous system.

menisci (men-is´si): wedge-shaped cartilages in certain synovial joints.

menopause (men´o-pawz): the cessation of menstrual periods in the human female.

menses (men´sez): the monthly flow of blood from the female genital tract.

menstrual cycle (men´stru-al): the rhythmic female reproductive cycle, characterized by changes in hormone levels and physical changes in the uterine lining.

menstruation (men´stru-a´shun): the discharge of blood and tissue from the uterus at the end of the menstrual cycle.

mesentery (mes´en-ter´e): a fold of peritoneal membrane that attaches an abdominal organ to the abdominal wall.

mesoderm (mes´o-derm): the middle one of the three primary germ layers.

mesothelium (mes´o-the´leum): a simple squamous epithelial tissue that lines body cavities and covers visceral organs; also called *serosa*.

metabolism (me-tab´o-lizm): the chemical changes that occur within a cell.

metacarpus (met´ah-kar´pus): the region of the hand between the wrist and the digits, including the five bones that support the palm of the hand.

metastasis (me-tas´tah-sis): the spread of a disease from one organ or body part to another.

metatarsus (met´ah-tar´sus): the region of the foot between the ankle and the digits, containing five bones.

microbiology (mi-kro-bi-ol´o-je): the science dealing with microscopic organisms, including bacteria, fungi, viruses, and protozoa.

microvilli (mi´kro-vil´i): microscopic, hairlike projections of cell membranes on certain epithelial cells.

midbrain: the portion of the brain between the pons and the forebrain.

middle ear: the middle of the three ear chambers, containing the three auditory ossicles.

mitosis (mi-to´sis): the process of cell division, in which the two daughter cells are identical and contain the same number of chromosomes.

mitral valve (mi´tral): the left atrioventricular heart valve; also called the bicuspid valve.

mixed nerve: a nerve containing both motor and sensory nerve fibers.

molecule (mol´e-kule): a minute mass of matter, composed of a combination of atoms that form a given chemical substance or compound.

motor neuron (nu´ron): a nerve cell that conducts action potential away from the central nervous system and innervates effector organs (muscles and glands); also called *efferent neuron*.

motor unit: a single motor neuron and the muscle fibers it innervates.

mucosa (mu-ko´sah): a mucous membrane that lines cavities and tracts opening to the exterior.

muscle (mus´el): an organ adapted to contract; three types of muscle tissue are cardiac, smooth, and skeletal.

mutation (mu-ta´shun): a variation in an inheritable characteristic, a permanent transmissible change in which the offspring differ from the parents.

myelin (me´e-lin): a lipoprotein material that forms a sheathlike covering around nerve fibers.

myocardium (mi´o-kar´de-um): the cardiac muscle layer of the heart.

myofibril (mi´o-fi´bril): a bundle of contractile fibers within muscle cells.

myoneural junction (mi´o-nu´ral): the site of contact between an axon of a motor neuron and a muscle fiber.

myosin (mi´o-sin): a thick filament protein that together with actin causes muscle contraction.

nail: a hardenend, keratinized plate that develops from the epidermis and forms a protective covering on the dorsal surfaces of the digits.

nares (na´rez): the opening into the nasal cavity; also called *nostrils*.

nasal cavity (na´zal): a mucosa-lined space above the oral cavity, which is divided by a nasal septum and is the first chamber of the respiratory system.

nasal septum (sep´tum): a bony and cartilaginous partition that separates the nasal cavity into two portions.

nephron (nef´ron): the functional unit of the kidney, consisting of a glomerulus, glomerular capsule, convoluted tubules, and the loop of the nephron.

nerve: a bundle of nerve fibers outside the central nervous system.

neurofibril node (nu´ro-fi´bril): a gap in the myelin sheath of a nerve fiber; also called the *node of Ranvier*.

neuroglia (nu-rog´le-ah): specialized supportive cells of the central nervous system.

neurolemmocyte (nu´ri-lem-o´site): a specialized neuroglia cell that surrounds an axon fiber of a peripheral neuron and forms the neurilemmal sheath; also called the *Schwann cell*.

neuron (nu´ron): the structural and functional unit of the nervous system, composed of a cell body, dendrites, and an axon; also called a *nerve cell*.

neutrophil (nu´tro-fil): a type of phagocytic white blood cell.

nipple: a dark pigmented, rounded projection at the tip of the breast.

node of Ranvier (rah-ve-a´): see *neurofibril node*.

notochord (no´to-kord): a flexible rod of tissue that extends the length of the back of an embryo.

nucleus (nu´kle-us): a spheroid body within a cell that contains the genetic factors of the cell.

nurse cells: specialized cells within the testes that supply nutrients to developing spermatozoa; also called *sertoli cells* or *sustentacular cells*.

olfactory (ol-fak´to-re): pertaining to the sense of smell.

oocyte (o´o-site): a developing egg cell.

oogenesis (o´o-hen´e-sis): the process of female gamete formation.

optic (op´tik): pertaining to the eye and the sense of vision.

optic chiasma (ki-as´mah): an X-shaped structure on the inferior aspect of the brain where there is a partial crossing over of fibers in the optic nerves.

optic disc: a small region of the retina where the fibers of the ganglion neurons exit from the eyeball to form the optic nerve; also called the *blind spot*.

oral: pertaining to the mouth; also called *buccal*.

organ: a structure consisting of two or more tissues, which performs a specific function.

organelle (or´gan-el´): a minute structure of a cell with a specific function.

organism (or´gah-nizm): an individual living creature.

orifice (or´i fis): an opening into a body cavity or tube.

origin (or´i-jin): the place of muscle attachment onto the more stationary point or proximal bone; opposite the insertion.

osmosis (os-mo´sis): the passage of a solvent, such as water, from a solution of lesser concentration to one of greater concentration through a semipermeable membrane.

ossicle (os´si-l´l): one of the three bones of the middle ear.

osteocyte (os´te-o-site): a mature bone cell.

osteon (os´te-on): a group of osteocytes and concentric lamellae surrounding a central canal within bone tissue; also called a *haversian system*.

ovarian follicle (o-va´re-an fol´li-k´l): a developing ovum and its surrounding epithelial cells.

ovary (o´vah-re): the female gonad in which ova and certain sexual hormones are produced.

oviduct (o´vi-dukt): the tube that transports ova from the ovary to the uterus; also called the *uterine tube* or *fallopian tube*.

ovulation (o´vu-la´shun): the rupture of an ovarian follicle with the release of an ovum.

ovum (o´vum): a secondary oocyte after ovulation but before fertilization.

palate (pal´at): the roof of the oral cavity.

palmar (pal´mar): pertaining to the palm of the hand.

pancreas (pah´kre-as): organ in the abdominal cavity that secretes gastric juices into the GI tract and insulin and glucagon into the blood.

pancreatic islets (pan´kre-at´ik): a cluster of cells within the pancreas that forms the endocrine portion of the pancreas; also called *islets of Langerhans*.

papillae (pah-pil´e): small nipplelike projections.

paranasal sinus (par´ah-na´zal si´nus): an air chamber lined with a mucous membrane that communicates with the nasal cavity.

parasympathetic (par´ah-smi´pah-thet´ik): pertaining to the division of the autonomic nervous system concerned with activities that are antagonistic to sympathetic.

parathyroids (par´ah-thi´roids): small endocrine glands that are embedded on the posterior surface of the thyroid glands and are concerned with calcium metabolism.

parietal (pah-ri´e-tal): pertaining to a wall of an organ or cavity.

parotid gland: (pah-rot´id): one of the paired salivary glands on the side of the face over the masseter muscle.

parturition (par´tu-rish´un): the process of childbirth.

pathogen (path´o-jen): any disease-producing organism.

pectoral girdle (pek´to-ral): the portion of the skeleton that supports the upper extremities.

pelvic (pel´vik): pertaining to the pelvis.

pelvic girdle: the portion of the skeleton to which the lower extremities are attached.

penis (pe´nis): the external male genital organ, through which urine passes during urination and which transports semen to the female during coitus.

pericardium (per´i-kar´de-um): a protective serous membrane that surrounds the heart.

perineum (per´i-ne´um): the floor of the pelvis, which is the region between the anus and the scrotum in the male and between the anus and the vulva in the female.

periosteum (per´e-os´te-um): a fibrous connective tissue covering the outer surface of bone.

peripheral nervous system (pe-rif´er-al): the nerves and ganglia of the nervous system that lie outside of the brain and spinal cord.

peristalsis (per´i-stal´sis): rhythmic contractions of smooth muscle in the walls of various tubular organs, which move the contents along.

peritoneum (per´i-to-ne´um): the serous membrane that lines the abdominal cavity and covers the abdominal viscera.

phagocyte (fag´o-site): any cell that engulfs other cells, including bacteria, or small foreign particles.

phalanx (fa´lanks), pl. *phalanges*: a bone of the finger or toe.

pharynx (far´inks): the organ of the GI tract and respiratory system located at the back of the oral and nasal cavities and extending to the larynx anteriorly and the esophagus posteriorly; also called the *throat*.

physiology (fiz´e-ol´o-je): the science that deals with the study of body functions.

pia mater (pi´ah ma´ter): the innermost meninx that is in direct contact with the brain and spinal cord.

pineal gland (pin´e-al): a small cone-shaped gland located in the roof of the third ventricle.

pituitary gland (pi-tu´i-tar´e): a small, pea-shaped endocrine gland situated on the inferior surface of the brain that secretes a number of hormones; also called the *hypophysis* and commonly called the *"master gland."*

placenta (plah-sen´tah): the organ of metabolic exchange between the mother and the fetus.

plasma (plaz´mah): the fluid, extracellular portion of circulating blood.

platelets (plate´lets): fragments of specific bone marrow cells that function in blood coagulation: also called *thrombocytes*.

pleural membranes (ploor´al): serous membranes that surround the lungs and line the thoracic cavity.

plexus (plek´sus): a network of enterlaced nerves or vessels.

plica circulares (pli´kah ser-ku-lar´is): a deep fold within the wall of the small intestine that increases the absorptive surface area.

pons (ponz): the portion of the brain stem just above the medulla oblongata and anterior to the cerebellum.

posterior (dorsal) (pos-ter´e-or): toward the back.

pregnancy: a condition where a female has a developing off-spring in the uterus.

prenatal (pre-na´tal): the period of offspring development during pregnancy; before birth.

proprioceptor (pro´pre-o-sep´tor): a sensory nerve ending that responds to changes in tension in a muscle or tendon.

prostate (pros´tate): a walnut-shaped gland surrounding the male urethra just below the urinary bladder that secretes an additive to seminal fluid during ejaculation.

proximal (prok´si-mal): closer to the midline of the body or origin of an appendage: opposite of *distal*.

puberty (pu´ber-te): the period of development in which the reproductive organs become functional.

pulmonary (pul´mo-ner´e): pertaining to the lungs.

pupil: the opening through the iris that permits light to enter the vitreous chamber of the eyeball and be refracted by the lens.

receptor (re-sep´tor): a sense organ or a specialized end of a sensory neuron that receives stimuli from the environment.

rectum (rek´tum): the portion of the GI tract between the sigmoid colon and the anal canal.

reflex arc: the basic conduction pathway through the nervous system, consisting of a sensory neuron, interneuron, and a motor neuron.

renal (re´nal): pertaining to the kidney.

renal corpuscle (kor´pus´l): the portion of the nephron consisting of the glomerulus and a glomerular capsule.

renal pelvis: the inner cavity of the kidney formed by the expanded ureter and into which the calyces open.

replication (re-pli-ka´shun): the process of producing a duplicate; a copying or duplication, such as DNA replication.

respiration: (res´pi-ra´shun): the exchange of gases between the external environment and the cells of an organism.

rete testis (re´te tes´tis): a network of ducts in the center of the testis, site of spermatozoa production.

retina (ret´i-nah): the inner layer of the eye that contains the rods and cones.

retraction (re-trak´shun): the movement of a body part, such as the mandible, backward on a plane parallel with the ground; the opposite of *protraction*.

rod: a photoreceptor in the retina of the eye that is specialized for colorless, dim light vision.

rotation (ro-ta´shun): the movement of a bone around its own longitudinal axis.

rugae (ru´je): the folds or ridges of the mucosa of an organ.

sagittal (saj´i-tal): a vertical plane through the body that divides it into right and left portions.

salivary gland (sal´i-ver-e): an accessory digestive gland that secretes saliva into the oral cavity.

sarcolemma (sar´ko-lem´ah): the cell membrane of a muscle fiber.

sarcomere (sar´ko-mere): the portion of a skeletal muscle fiber between the two adjacent Z lines that is considered the func-

tional unit of a myofibril.

Schwann cell (shwahn): see *neurolemmocyte*.

sclera (skle´rah): the outer white layer of connective tissue that forms the protective covering of the eye.

scrotum (skro´tum): a pouch of skin that contains the testes and their accessory organs.

sebaceous gland (se-ba´shus): an exocrine gland of the skin that secretes *sebum*, an oily protective product.

semen (se´men): the secretion of the reproductive organs of the male, consisting of spermatozoa and additives.

semicircular ducts: tubular channels within the inner ear that contain the receptors for equilibrium; also called *semicircular canals*.

semilunar valve (sem´e-lu´nar): crescent-shaped heart valves, positioned at the entrances to the aorta and the pulmonary trunk.

seminal vesicles (sem´i-nal ves´i-k´lz): a pair of accessory male reproductive organs lying posterior and inferior to the urinary bladder, which secrete additives to spermatozoa into the ejaculatory ducts.

sensory neuron (nu´ron): a nerve cell that conducts an impulse from a receptor organ to the central nervous system; also called *afferent neuron*.

serous membrane (se´rus): an epithelial and connective tissue membrane that lines body cavities and covers viscera; also called *serosa*.

sesamoid bone (ses´ah-moid): a membranous bone formed in a tendon in response to joint stress.

sigmoid colon (sig´moid ko´lon): the S-shaped portion of the large intestine between the descending colon and the rectum.

sinoatrial node (sin´o-a´tre-al): a mass of cardiac tissue in the wall of the right atrium that initiates the cardiac cycle; the SA node; also called the *pacemaker*.

sinus (si´nus): a cavity or hollow space within a body organ such as a bone.

skeletal muscle: a type of muscle tissue that is multinucleated, occurs in bundles, has crossbands of proteins, and contracts either in a voluntary or involuntary fashion.

small intestine: the portion of the GI tract between the stomach and the cecum, functions in absorption of food nutrients.

smooth muscle: a type of muscle tissue that is nonstriated, composed of fusiform, single-nucleated fibers, and contracts in an involuntary, rhythmic fashion within the walls of visceral organs.

somatic (so-mat´ik): pertaining to the nonvisceral parts of the body.

spermatic cord (sper´mat´ik): the structure of the male reproductive system composed of the ductus deferens, spermatic vessels, nerve, cremasteric muscle, and connective tissue.

spermatogenesis (sper´mah-to-jen´e-sis): the production of male sex gametes, or spermatozoa.

spermatozoon (sper´mah-to-zo´on): a sperm cell, or gamete.

sphincter (sfingk´ter): a circular muscle that constricts a body opening or the lumen of a tubular structure.

spinal cord (spi'nal): the portion of the central nervous system that extends from the brain stem through the vertebral canal.

spinal nerve: one of the thirty-one pairs of nerves that arise from the spinal cord.

spleen: a large, blood-filled organ located in the upper left of the abdomen and attached by the mesenteries to the stomach.

spongy bone (spun'je): a type of bone that contains many porous spaces; also called *cancellous bone.*

stomach: a pouchlike digestive organ between the esophagus and the duodenum.

submucosa (sub'mu-ko'sah): a layer of supportive connective tissue that underlies a mucous membrane.

superior vena cava (ve'nah ka'vah): a large systemic vein that collects blood from regions of the body superior to the heart and returns it to the right atrium.

surfactant (ser-fak'tant): a substance produced by the lungs that decreases the surface tension within the alveoli.

suture (su'chur): a type of fibrous joint articulating between bones of the skull.

sympathetic (sim'pah-thet'ik): pertaining to that part of the autonomic nervous system concerned with activities antagonistic to the parasympathetic.

synapse (sin'aps): a minute space between the axon terminal of a presynaptic neuron and a dendrite of a postsynaptic neuron.

synovial cavity (si-no've-al): a space between the two bones of a synovial joint, filled with synovial fluid.

system: a group of body organs that function together.

systole (sis'to-le): the muscular contraction of the ventricles of the heart during the cardiac cycle.

systolic pressure (sis'tol'ik): arterial blood pressure during the ventricular systolic phase of the cardiac cycle.

target organ: the specific body organ that a particular hormone affects.

tarsus (tahr'sus): pertaining to the ankle; the proximal portion of the foot that contains the seven tarsal bones.

tendo calcaneous (ten'do kal-ka'ne-us): the tendon that attaches the calf muscles to the calcaneous bone.

tendon (ten'dun): a band of dense regular connective tissue that attaches muscle to bone.

testis (tes'tis): the primary reproductive organ of a male, which produces spermatozoa and male sex hormones.

thoracic (tho-ras'ik): pertaining to the chest region.

thoracic duct: the major lymphatic vessel of the body, which drains lymph from the entire body except the upper right quadrant and returns it to the left subclavian vein.

thorax (tho'raks): the chest.

thymus gland (thi'mus): a bi-lobed lymphoid organ positioned in the upper mediastinum, posterior to the sternum and between the lungs.

tissue: an aggregation of similar cells and their binding intercellular substance, joined to perform a specific function.

tongue: a protrusible muscular organ on the floor of the oral cavity.

trachea (tra'ke-ah): the airway leading from the larynx to the bronchi; also called the *windpipe.*

tract: a bundle of nerve fibers within the central nervous system.

transverse colon (ko'lon): a portion of the large intestine that extends from right to left across the abdomen between the hepatic and splenic flexures.

tricuspid valve (tri-kus'pid): the heart valve between the right atrium and the right ventricle.

tympanic membrane (tim-pan'ik): the membranous eardrum positioned between the outer and middle ear; also called the *tympanum,* or the *ear drum.*

umbilical cord (um-bil'i-kal): a cordlike structure containing the umbilical arteries and vein, which connects the fetus with the placenta.

umbilicus (um-bil'i-kus): the site where the umbilical cord was attached to the fetus: also called the *navel.*

ureter (u-re'ter): a tube that transports urine from the kidney to the urinary bladder.

urethra (u-re'thrah): a tube that transports urine from the urinary bladder to the outside of the body.

urinary bladder (u're-ner'e): a distensible sac in the pelvic cavity which stores urine.

uterine tube (u'ter-in): the tube through which the ovum is transported to the uterus and where fertilization takes place: also called the *oviduct* or *fallopian tube.*

uterus (u'ter-us): a hollow, muscular organ in which a fetus develops. It is located within the female pelvis between the urinary bladder and the rectum.

uvula (u'vu-lah): a fleshy, pendulous portion of the soft palate that blocks the nasopharynx during swallowing.

vagina (vah-ji'nah): a tubular organ that leads from the uterus to the vestibule of the female reproductive tract and receives the male penis during coitus.

vein: a blood vessel that conveys blood toward the heart.

ventral (ven'tral): toward the front surface of the body: also called *anterior.*

vestibular folds: the supporting folds of tissue for the vocal folds within the larynx.

vestibular window: a membrane-covered opening in the bony wall between the middle and inner ear, into which the footplate of the stapes fits; also called *oval window.*

viscera (vis'er-ah): the organs within the abdominal or thoracic cavities.

vitreous humor (vit're-us hu'mor): the transparent gell that occupies the space between the lens and retina of the eye.

vocal folds: folds of the mucous membrane in the larynx that produce sound as they are pulled taut and vibrated; also called *vocal cords.*

vulva (vul'vah): the external genitalia of the female that surround the opening of the vagina; also called the *pudendum.*

zygote (zi'gote): a fertilized egg cell formed by the union of a sperm and an ovum.

Index